Green Development or Greenwashing?

Green Development or Greenwashing?

Environmental Histories of Finland

edited by

Viktor Pál, Tuomas Räsänen and Mikko Saikku

The White Horse Press

The White Horse Press, The Old Vicarage, Main Street, Winwick, Cambridgeshire PE28 5PN, UK

Set in 11 point Adobe Garamond Pro

doi: 10.3197/63824846758018.book

British Library Cataloguing in Publication Data
A catalogue record for this book is available from the British Library

ISBN: 978-1-912186-76-1 (PB); 978-1-912186-77-8 (Open Access E-book)

CONTENTS

CONTRIBUTOR BIOGRAPHIES

Editors

Viktor Pál is a Hungarian environmental historian, an associate professor at the University of Tampere and the University of Ostrava, and a visiting researcher at the University of Helsinki. He is the author of *Technology and the Environment in State-socialist Hungary: An Economic History* (Palgrave Macmillan, 2017) and with Stephen Brain has co-edited the collection of essays, *Environmentalism under Authoritarian Regimes. Myth, Propaganda, Reality* (Routledge, 2019).
Email: viktor.paal@gmail.com

Tuomas Räsänen works as an associate professor of environmental history at the University of Eastern Finland. His research interests include the history of human-wild animal relationship, the history of Finnish environmentalism and the Baltic Sea marine environmental history.
Email: tuomas.rasanen@uef.fi

Mikko Saikku is the McDonnell Douglas Professor of American Studies at the University of Helsinki. He is the author or editor of several internationally noted academic articles, collections and books, including *This Delta, This Land: An Environmental History of the Yazoo-Mississippi Delta* (University of Georgia Press, 2005) and *An Unfamiliar America: Essays in American Studies* (Routledge, 2021). He is past president of the Nordic Association for American Studies, and currently serves as Director of the Helsinki Environmental Humanities Hub.
Email: mikko.saikku@helsinki.fi

Authors

Matti O. Hannikainen works as a post-doctoral researcher at the University of Helsinki. He is the author of *Greening of London 1920–2000* (Ashgate 2016). He has specialised in environmental and urban history. His current focus is on cultural history of fish in Finnish society from the 1850s to the present.
Email: matti.o.hannikainen@helsinki.fi

Oona Ilmolahti has a Ph.D. in history, and she is a project researcher at the University of Eastern Finland. Her interests lie in the history of emotions and senses, post-war societies, transgenerational trauma, museology and culture nature interaction. In recent years she has studied border regions, in particular Karelian identities, as well as biographies. Her latest article concerns the multisensory history of Lake Ladoga (2022).
Email: oona.ilmolahti@gmail.com
https://orcid.org/0000-0003-2914-2128

Simo Laakkonen (Dr.Sc.Soc.) is director of the Degree Program in Digital Culture, Landscape, and Cultural Heritage Studies at the University of Turku, Finland. He has explored the environmental history of the Baltic Sea as well as of World War II and the Cold War. Recently, he has edited two books: *The Long Shadows: A Global Environmental History of the Second World War* and *The Resilient City in World War II: Urban Environmental Histories.*
Email: sijula@utu.fi

Heta Lähdesmäki is a historian specialising in human-animal studies, human-wildlife conflicts and conservation. She completed a Ph.D. in cultural history in 2020 at the University of Turku, studying human-wolf relations in twentieth century Finland. After that, she studied the relationship between humans and nature in Seili island in a multidisciplinary research project, Seili - Elämän saari, funded by the Kone Foundation and led by the Biodiversity unit at the University of Turku. She is part of the Academy of Finland funded HumBio-project, investigating the human relationship with disappeared, endangered, introduced and non-native, as well as invasive, marine animals and plants in Finland. Currently, Lähdesmäki is a postdoctoral researcher at the University of Helsinki and the Helsinki Institute of Sustainability Science. She is part of the Helsinki Urban Rat Project and studies the history of bird feeding and rat conflicts in Helsinki city.
Email: heta.lahdesmaki@helsinki.fi

Maria Lähteenmäki is Professor in History at the University of Eastern Finland and Adjunct Professor at the University of Helsinki. She specialises in the history of sub-arctic border regions and communities such as Lapland, the Barents region and Karelia. She has published works such as *The Peoples in Lapland. Borders and Interaction in the North Calotte 1808–1889* (2006) and *Footprints in the Snow: A Long History of the Arctic Finland* (2017). She co-edited *Lake*

Ladoga. The Coastal History of the Greatest Lake in Europe (2023) and co-edited *The Barents Region. A Transnational History of Subarctic Northern Europe* (2015). Her latest work is *Punapakolaiset*, which chronicles Finnish red refugees in Soviet Karelia in 1918–1938.
Email: maria.lahteenmaki@uef.fi
https://orcid.org/0000-0001-9142-659X

Jaana Laine is a university lecturer in social sciences at the LUT University (Lappeenranta, Finland) and an associate professor at the University of Helsinki. Her publications concentrate on Finnish forest history, covering, for instance, timber trade institutions, forest workers and labour markets and the history of the Finnish Forest Research Institute. Recent research interests include human-forest relationships.
Email: Jaana.Laine@lut.fi

Janne Mäkiranta works as a post-doctoral researcher at the University of Turku in the department of European and World History. His research focuses on the entanglements of history of science and environmental history.
Email: jajuma@utu.fi

Outi Manninen is a Doctor of Natural Sciences and Senior researcher. She is a plant ecologist, currently working as a researcher at the University of Lapland and Natural Resources Institute Finland. She is a member of the project 'Historical sites as a novel tool for predicting long-term boreal and subarctic ecosystem change' (HISTECO). Her latest co-authored article concerns historical reindeer corrals as portraits of human-nature relationships in Northern Finland (2022).
Email: outi.manninen@ulapland.fi
https://orcid.org/0000-0002-8438-2039

Risto-Matti Matero (MA, M.Soc.Sci) is a Ph.D. Candidate in General History in the Department of History and Ethnology at the University of Jyväskylä. He is currently writing his dissertation on the development of environmental ideas in the 1980s and 1990s in Finnish and German Green Parties.
Email: rmatero85@gmail.com

Seija A. Niemi is an independent scholar. Her doctoral thesis (2018) discusses the Finnish-Swedish explorer and scientist Adolf Erik Nordenskiöld (1832–1901) and his place in the history of early Nordic conservation. Her publications

include, for example, 'Exploring environmental literacy from a historical perspective', in Estelita Vaz, Cristina Joanaz de Melo and Ligia M. Costa Pinto (eds), *Environmental History in the Making. Volume I: Explaining* (New York: Springer, 2016) and 'How fossils gave the first hints of climate change', in Dolly Jörgensen and Sverker Sörlin (eds), *Northscapes: History, Technology, and the Making of Northern Environments* (Vancouver: University of British Columbia Press, 2013). Niemi's other articles can be found in several scientific journals.
Email: seija.astrid@gmail.com

Jukka Nyyssönen works as a Research Professor at The High North Department at the Norwegian Institute for Cultural Heritage Research in Tromsø, Norway. He has expertise on Sámi history, which he has studied from numerous perspectives, including but not limited to Environmental History, History of Minorities and Animal History.
Email: Jukka.Nyyssoenen@niku.no

Mauri Soikkanen, MA, is a retired journalist, editor and historian. Soikkanen studied natural history and biology at the University of Helsinki and worked as a geography assistant after graduation. He started as the first nature editor of the Finnish Broadcasting Corporation in 1956, later becoming the programme manager. Soikkanen was the editor-in-chief of Finland's longest-running hunting and fishing magazine. He has written or edited over thirty books on the history of hunting and fishing, including one book on World War Two.

Sari Stark is Docent in Plant Ecology and university researcher in the University of Lapland. Her research focus is the effects of global changes on northern ecosystems. She is the PI of the project 'Historical sites as a novel tool for predicting long-term boreal and subarctic ecosystem change' (HISTECO), financed by Academy of Finland 2019–2023. Her latest co-authored article is 'The ecosystem effects of reindeer (Rangifer tarandus) in northern Fennoscandia: Past, present and future', *Perspectives in Plant Ecology, Evolution and Systematics* **58** (2023).
Email: sari.stark@ulapland.fi
https://orcid.org/0000-0003-4845-6536

ABSTRACTS

Chapter 2. Knowledge of Trees and Forests – Finnish Forest Research from the Nineteenth to the Twentieth Century

Jaana Laine

Finnish forestry and forest science reflect demands set by the state administration and the forest industry but also private forest and nature conservation organisations, and nowadays private citizens e.g., through social media. From the late nineteenth century to the 2020s, the history of forests, forest science and Finnish society consists of four main periods.

During the first period – *know the forests* (late nineteenth century–1930s) – society needed and gained information on forests, especially on growing timber stock (the first forest inventory in the 1920s) and wood consumption (the first inventory of wood consumption in the 1930s). In addition, researchers produced knowledge for silvicultural practices and forest biology.

Rationalising forestry and developing timber procurement were seen as essential during the second period – *exploit the forests* (1940s–1960s). Since timber removals exceeded annual growth, the state launched massive forest improvement actions. Large clear-cuttings were regenerated with conifer saplings and massive draining of bogs was enacted. As a result, society more extensively exploited and influenced the forests.

During the third period – *define the forests* (1970s–1990s) –forests were no longer respected merely as a source of economic prosperity. Escalating disputes came about when environmental activism and public discussions challenged forestry practices. Scientific knowledge became imbricated, besides traditional forestry values, also with nature conservation, recreational and environmental values related to forests. During the 1990s, changes in forest legislation paved the way for more pluralistic values of forests.

During the most recent period – *discover forests' futures* (2000s–) – climate change and conflicting human-forest relationships set new demands for forestry and forest science. Forests' importance has grown from the private and national level to the global context. Forests are respected as carbon sinks and storage, for their rich biodiversity, and as a source of mental and physical health. Forests as living entities are often recognised and new steps have been taken towards more pluralistic human-forest relationships, posthumanism and interspecies perspectives.

doi: 10.3197/63824846758018.abstracts

Chapter 3. 'Reaching Maturity' or 'Selling Out'? The Idea of Green Growth in Finnish Green Party Environmental Discourses 1988–1995.

Risto-Matti Matero

Over the past decades, major shifts have taken place in public environmental discourses transnationally, of which the Finnish Green party provides an illustrative example. Green parties were formed throughout Europe to represent radical alternative social movements and their growth-critical ideals. By the turn of the millennium, however, earlier radicalism was transformed into moderate ideals of green growth.

This chapter demonstrates how green growth ideals were used as a political tool by the Finnish Green Party to better adapt to a free market political system, as well as some of the premises with which this turn was implemented. As a political act, the goal of implementing green growth ideals was to be more efficient within the prevailing political system. The need for such pragmatism can be explained with William Connolly's framework of *cultural belonging:* in order to act meaningfully, one needs to adapt to the premises of the culture one operates in, causing a challenge for paradigm-shifting environmentalism to become implemented politically. The case of Finnish Green party ideological development provides an example of this transnational phenomenon.

Chapter 4. The Changing Status of Birch Trees in Finnish Forests from the Seventeenth Century to the Twentieth Century

Seija A. Neimi

This chapter illuminates the changing status of the birch tree, how the Finns have perceived it, and what have Finnish standards of environmental literacy have been from the seventeenth century to the twentieth century. This period covers both pre-industrial and industrial socio-economic changes, from the ancient hunters and slash-and-burn cultivators to modern architecture, art and wood processing industries. Finland's forests are relatively the largest in Europe: 86 % of the country's surface is covered with the woods. The three most common tree species are pine, spruce and birch. The value of pine and spruce grew significantly when the wood processing industry began to use wood fibres in production, while birch has had its ups and downs which makes it an interesting tree to study.

Chapter 5. Trash Food? Fish as Food in Finnish Society Between the 1870s and the 1990s

Matti O. Hannikainen

The relationship between the Finns and fish as food changed drastically during the twentieth century. In this chapter, we shall explore how the concept 'trash fish', which refers to those species with little or no value for human consumption, was invented and how it evolved and affected the consumption of fish in Finnish society. A scientific discourse aimed at rationalising fishing by classifying fish species according to their (potential) commercial value, thus promoting the valuable species and labelling a few as trash. More importantly, Finns began to prefer both fresh and imported frozen fish over salted fish. This had a drastic impact on the consumption of fish, marking a change captured in numerous cookbooks. Based on the textual analysis of official documents, fishing manuals, journal articles and cookbooks all published in Finnish, we will explore how value of various fish species reflected changes in scientific and culinary discourses.

Chapter 6. Cultural Nature in Mid-Lappish Reindeer Herding Communities

Maria Lähteenmäki, Oona Ilmolahti, Outi Manninen and Sari Stark

Our research task is to present and analyse features of the local human-nature and human-reindeer relations in the historical timespan of the twentieth century and in the context of cultural nature in the historical Forest Sámi area of Finnish Mid-Lapland. By *cultural nature* we refer to the different meanings and attributes groups and individuals give and have given to their surrounding natural environment with its fauna, flora, and waterways. The question is viewed through environmental changes and the meanings connected to reindeer roundups (corrals) and roundup places as an example of human-nature interaction. The reindeer roundups have historically been important social meeting places for subarctic communities, and the roundup events have traditionally been the highlight of the reindeer year. Our empirical focus lies in two *reindeer herding cooperatives* (Finn. *paliskunta*), Sattasniemi and Oraniemi, geographically located in the middle of Finnish Lapland – mainly in Sodankylä, and partly in Savukoski and Pelkosenniemi municipalities – and the reindeer roundup processes in these cooperatives. Our key source data consists of archival material, such as the minutes and reports of the Reindeer Herders Association and Sattasniemi co-operative. We have also utilised regional, local and occupational newspapers and magazines from the 1920s to the 2010s. In order to reach the voices of the contemporary herder communities we conducted a *Cultural Nature Survey* from 22 February to 30 March 2021. In the course of the twentieth century, Mid-Lapland faced enormous environmental changes. Intensive forestry, energy production and the mining industry have physically altered the landscape and disturbed reindeer herding based on natural pasture rotation. Continuity of livelihood and way of life are worrying issues in the region. The feeling of not being heard or understood also affects communities' nature and reindeer relationships. The more the surrounding natural and cultural environments have changed, the more the Mid-Lappish communities have tried to revitalise the 'original' nature-human-reindeer relationship and the nostalgic stories of dense forests, free waterways and untouched wilderness. The locals emphasise their 'authentic' Lappish lifestyle, at least in terms of reindeer herding. This endeavour can be regarded as cultural use of nature.

The article was prepared in cooperation between the University of Lapland and the University of Eastern Finland. It is part of the HISTECO project (2019–2023) funded by the Academy of Finland and led by Sari Stark.

Chapter 7. Sámi Frames in the Planning and Management of Nature Protection Areas in Historical Perspective – Environmental Non-conflict in Inari

Jukka Nyyssönen

What kind of framings can be detected in Sámi opinions on conservation of nature in Inari? The region has witnessed recurrent conflicts over land usage, fought between forestry officials and Sámi herders. Establishment of nature reserves has aroused severe disputes as well, but conservation enjoys continuing support among the Sámi herders. This article charts the preconditions for this state of affairs through cases of the establishment of the state park of Koilliskaira (1975-1982) and recent administrative measures in park administration by the Sámi Parliament (2000s). An analysis is undertaken of whether the frames concerning conservation aligned in the administrative setting and the background reasons for the (non-)alignment. The actors studied are those Sámi included in the establishment processes and the park administration: the Sámi herders and the Sámi Parliament. The conservation history is contextualised in the history of the Sámi movement and its relations to state actors, the Forest and Park Service (FPS). The case is one of success for both conservationists and Sámi. The Sámi mostly favoured conservation, because the protection of parks meant protection of reindeer herding from competing land-use forms. Later, conservation became a way to manifest the cultural autonomy, self-determination and cultural rights of the Sámi. An institutional source for this success was the marginalisation of the FPS from park establishment processes. The case was framed mostly economically, as a possibility to safeguard the pastures from forestry, and later as a case of indigenous rights. The economic framing resonated well both with conservationists and the general sentiments of the era; only later did indigenous rights clash with environmental values.

Chapter 8. Wolves and the Finnish Wilderness: Changing Forests and the Proper Place for Wolves in Twentieth-Century Finland

Heta Lähdesmäki

These days, if wolves roam close to human settlements, people often argue that there is something problematic and unnatural in it. This is the case especially in western Finland where wolf packs are being observed after a long period of absence. Not everyone living in western Finland has welcomed wolves as neighbours. Local people can argue that wolves should not live in western Finland because there are no wilderness areas there. Wolves have been connected to the wilderness in many countries and regions in the world. In some areas, the notion that the wolf belongs to the wilderness is old:

For instance, historian Aleksander Pluskowski has argued that there was a persistent conceptual link between wolves and the wilderness in Britain and Scandinavia during the Middle Ages. In this chapter, I look into this notion and trace its history in the Finnish context by studying newspaper reports, magazine articles and contemporary literature. I argue that the idea that wolves belong to the wilderness is a relatively new and controversial notion connected to various social and environmental changes. Interestingly, at the same time as the idea that the wolf is a wilderness species strengthened, the Finnish environment underwent changes that meant that the areas that could be called wilderness became fewer.

Chapter 9. All Quiet on the Eastern Front? The Finnish Army and Wildlife during World War Two

Mauri Soikkanen and Simo Laakkonen

World War II was arguably the most important event of the last century. The war was waged on almost all continents. It engulfed three-quarters of the world's population and was the world's most destructive war, claiming fifty to seventy million lives. In addition, the war injured hundreds of millions of people and innumerable other living creatures. Environmental history is a new way to explore the largest war that has taken place on Earth so far. This article focuses on the mobilisation of natural resources in this war. What role did hunting and fishing have during the Second World War? Depictions do exist in the memoirs of soldiers and officers from different countries, but hardly any historical studies have been conducted on these themes to date. This chapter is probably the first attempt made internationally to review the extent of hunting and fishing activity in wartime, and its importance both for military personnel and wildlife populations. It examines hunting and fishing along the thousand-kilometre border between Finland and the Soviet Union during World War Two. It focuses on the largest wilderness area in Europe where Finnish soldiers faced an oasis of wildlife that had disappeared from their own homesteads. However, within a short time period, this newly found wartime oasis disappeared too.

Chapter 10. From Stale Air to Toxic: Concerns About Urban Air in Finland

Janne Mäkiranta

Air pollution can be seen as one of the oldest and most enduring environmental problems in human history. At the same time its modern form as a public health concern is relatively novel. The purpose of this chapter is to examine the concern about urban air quality in Finland from the late nineteenth century to the late 1960s, when the fear of air pollution rose to new prominence. The chapter shows how nineteenth and early twentieth century concerns over urban air were based on the idea that clean air is beneficial for health. During the twentieth century, this vague hygienic idea was replaced by a more specific view that derived from medical research in industrial environments.

From the late 1950s onwards, air pollution measurements and medical research provided a detailed analysis of air quality. As a result, the concern about urban air was directed towards specific pollutants and their potential effects on health. This toxicological approach has been criticised in environmental history for its reductionism. The chapter shows, however, how the same approach was embraced by the environmental critics of the late 1960s. These critics used the toxicological approach to make local air pollution part of the concern about global environmental contamination. The chapter concludes that, despite the new rhetoric of the 1960s, the concern over urban air was in many ways a continuation of the nineteenth century critique about the hazards of urban living and the general progress of civilisation. The longing for clean air was replaced by the ubiquitous threat of toxins in the environment.

Chapter 11. From Eradication Campaigns to 'Care Protection': Finnish Endangered Animals in the Twentieth Century

Tuomas Räsänen

This chapter examines the human relationship with wild animals in Finland during the twentieth century. The chapter analyses three distinctive, yet somewhat overlapping, stages of human-animal relations. The first stage covers much of the first part of the century, when wild animals were perceived almost solely through the prism of their utility to humans. Game animals were considered as resources to be exploited, while predators were feared for the harm they might cause to humans and their domestic animals. As a result, many species from both categories were hunted to the brink of extinction. The second stage, around mid-century, saw the evolution of a more complex relation, when some species that were formerly hunted relentlessly were given protected status. Often these protected species were constructed as having cultural and historical significance for the Finnish people and thus being symbols of Finnishness. The third stage extends from the 1960s to the end of the century and beyond. During this stage, the protection of species and their habitats emerged as an elemental part of environmental discourse, with new labour-intensive techniques to protect wild animals. Yet, more than one tenth of Finnish animal species and half of habitats are endangered, and these trends have shown continuous deterioration.

Chapter 1

A 'GREEN SUPERPOWER'? INTRODUCTION TO THE ENVIRONMENTAL HISTORIES OF FINLAND

Viktor Pál, Tuomas Räsänen and Mikko Saikku

Finland has been often labelled a 'green superpower'. In 2016, according to the EPI (Environmental Performance Index) prepared by Yale and Columbia Universities, Finland was the world's cleanest and greenest country.[1] Generally speaking, Nordic countries have tended to be idealised as 'pristine and green' compared to the rest of the rapidly contaminating world where the race for markets and profits has generated an accelerated level of consumption.[2] Environmental historians, however, can detect that the commonly perceived 'greenness' of the Nordic countries is partly an illusion. One of the most notable examples of histories in this vein is the recent volume by Peder Anker, who interprets Norway's green development in the light of the country's former peripheral position and shows how Norway has become a global green leader, while developing a robust petroleum industry and having some of the worst CO_2 emissions per capita in the world.[3]

This volume states that Finland, like Norway, has evolved into being a green superpower at the price of considerable environmental problems: the current leadership position of Finland in sustainable development has been built on the heavy use of natural resources and at the expense of ecosystem health. Consequently, development and profit maximisation have had a significant and long-lasting negative impact on the natural environment in and around

1. Socioeconomic Data and Applications Center: https://sedac.ciesin.columbia.edu/data/collection/epi/sets/browse
2. For contamination and its causes worldwide, see McNeill 2001; Jarridge and Le Roux 2021.
3. Anker 2020.

doi: 10.3197/63824846758018.ch01

Finland. Old-growth forests have been replaced by intensive forest farming for lumber and pulp industries; more than half of wetlands have been drained for agriculture, forest cultivation and peat extraction; wild animal populations have been decimated; and Finland today is confined to the south and west by arguably one of the most polluted seas in the world.

The environmental harms of Finland's development have been widely studied in a number of sustainability sciences. For example, regional environmental sciences approaches have sought to understand the societal reasons behind the pollution of the Baltic Sea,[4] and criticised the newly emerged consumer lifestyle and massive material footprints of members of Finnish society,[5] as well as cases where Finnish industry is globally involved in environmentally controversial issues – such as UPM's involvement in Paso de los Toros in Uruguay, the world's largest pulp mill.[6]

In Finnish historiography, however, the dominant narrative has portrayed Finland as an economic miracle built upon the country's vast forest resources. This view may be valid when looking strictly from the human perspective of economic sustainability. Throughout the twentieth century, the wood processing industry was Finland's largest import sector and source of foreign capital, while scientifically managed forestry enabled Finnish forest reserves to actually increase. Not least by exploiting forests, Finland developed from being one of the poorest countries of Europe at the start of the twentieth century into a prosperous welfare state.[7] This storyline, however, ignores the enormous impact this 'economic miracle' has had on the natural environment, with old-growth forests replaced by monocultures that were often planted on drained wetlands.

Pursuing this vein, authors contributing to this volume aim to take a pioneering path and argue that a complex set of social-economic-technological as well as environmental factors helped Finland to build an image as an eco-leader nation while retaining its extractive economic practices to develop first industrial, then post-industrial, capitalism. The authors suggest that, partly due to the harsh climatic conditions of the Northern, subarctic environments, as well as to a long history of economic and technological backwardness, the conditions and implications of the climate, forests and water resources have been some of the main subjects of discussion in Finnish scientific and cultural discourses since the onset of enlightened thinking in Finland in the eighteenth

4. Hasler et al. 2019..
5. Lettenmeier et al. 2014.
6. Friends of the Earth Finland 2017.
7. E.g. Jalava et al. 2006; Kuisma 2006.

century. This trend expanded by the 1960s–1980s, during a period when Finland enjoyed continuous and unprecedented economic growth, albeit with a heavy toll on environmental quality.[8]

Consequently, we aim to re-evaluate the history of human-nature interactions in Finland, not only as a success-story of wise-use of the environment and the progress this entailed, but also in light of the country's high CO_2 emissions and ecological impact, half of which are generated by economic activity abroad. Instead of presenting a bucolic utopia, this volume aims to place some key aspects of Finnish environmental history under critical scrutiny.

The essays collected here suggest that the environmental history of Finland cannot be detached from global tendencies and an increasingly globalised world. Although the contributions cover diverse themes and introduce a wide range of aspects of human-nature relations, including the Sámi and Sápmi, forests, animals, pollutions, marine environment and politics, they are focused on key, globally relevant, layers of anthropogenic impact.

The chapters in Section 1 discuss the human perception and construction of the environment with a focus on Finland, beginning with Chapter 2, in which Jaana Laine analyses how the aims and activities of forestry and forest science reflected demands set by the state administration and forest industry, but also those of private forest and nature conservation organisations, as well as private citizens' opinions, as represented, for example, in social media. To do so, Laine uses a *longue durée* approach, starting in the late nineteenth century and spanning to the 2020s, identifying four different and formative time periods. According to Laine, during the first period, between the late nineteenth century and the 1930s, Finnish science and state administration aimed to collect information on forests, especially about growing timber stock, which resulted in the first comprehensive Finnish forest inventory project in the 1920s. That increased knowledge and understanding contributed to increased wood consumption and exploitation of the forests by the late 1940s. Science and state actors promoted reforestation and widespread land reclamation projects to improve the timber removal and annual growth balance, from an economic perspective, during the coming decades. Laine argues that, from the 1970s, increased environmental concerns were included in public discussions about forestry, partly due to the growing awareness of conservation, recreation and other environmental values in forest science knowledge. Environmental discussions widened after the 2000s, when climate change and conflicting human-forest relationships posed new demands for Finnish forestry and forest science.

8. Winiwarter et. al. 2004.

In the next chapter, Risto-Matti Matero analyses the rise of the concept of 'green growth' in a transnational environmental political debate, using the Finnish Green Party as a case study of how and why the concept was adopted. Matero argues that, during the 1990s, radical green visions disappeared from the Finnish Greens' political discussions. This led to the green party adopting notions of economic growth and consumerism. Via the new mainstream party paradigm, consumers would affect the environment by making positive choices in free markets, thus creating green growth in the globalising world. From the perspective of the Finnish Greens, this ideological change was interpreted as an attempt to better adapt environmental politics into the political and economic paradigm of the country. Employing the Finnish case, Matero's paper asks uncomfortable questions about the ideologies, goals and roles of environmentalist parties.

In Chapter 4, Seija A. Niemi investigates changing perceptions of birch trees in Finnish society over the past four centuries. Niemi argues that the economic importance of pine and spruce forests grew significantly because of the increased presence of the wood processing industry in Finland, whereas the birch tree experienced a different history. Niemi places the birch in socio-economic context, and examines the two-way, interspecies interactions governed on the human side by inherited language, attitudes and values. The author argues that the birch tree's marginal economic importance led to a more holistic understanding of the tree and contributed to Finns' high level of environmental literacy.

In the following chapter, Matti O. Hannikainen examines the classification of fish in Finnish society, and argues that the relationship between Finns and fish, as a food source, changed dramatically during the twentieth century. Over the past century some species of fish came to be labelled as 'trash fish' and Finnish consumers began to prefer both fresh and imported frozen fish, which drastically decreased the consumption of salted fish. Hannikainen explores how the concept 'trash fish', referring to species with little or no value for human consumption, gained prominence and in return affected the consumption of fish in Finnish society. The author uses textual analysis of official documents, fishing manuals, professional journal articles and cookbooks and pays particular attention to past scientific discourses that aimed at rationalising fishing by classifying species according to their commercial value.

Section 2 focuses on contested and colonised spaces and begins with Chapter 6 in which Maria Lähteenmäki, Oona Ilmolahti, Outi Manninen and Sari Stark investigate traditions of reindeer husbandry in Mid-Lapland via challenging the stereotypical images of Lapland. In the traditional home areas of the Forest Sámi, Lapland has undergone far-reaching environmental changes during the past few hundred years. Alongside environmental changes, that part of Lapland also underwent ethnic transition during the eighteenth and nineteenth centuries when Finnish settlers came into close contact with the Forest Sámi people. The Mid-Lapland area has been often seen by ethnic and sociocultural scholars as an uninteresting borderland. In opposition to that discussion, Lähteenmäki and Ilmolahti argue that the home of the Forest Sámi is a peculiar cultural area, which can be identified as the Mid-Lappish *hybrid cultural region*. That new notion may open fruitful research streams in the study of human–nature relationships.

In the next chapter Jukka Nyyssönen uncovers the recent environmental history of the multi-ethnic Inari area by analysing repeated disputes over resource use and the fate of reindeer-herding, in a context where logging and other land use forms have competed for the same areas. In the Inari area, expanding nature conservation areas have aroused local protests over feared restrictions to hunting rights and possibilities of logging the forest reserves. Interestingly, the Sámi herders remained positive towards the protected areas. Nyyssönen uses frame analysis to study the planning processes of the Lemmenjoki and Koilliskaira/Urho Kekkonen National Parks which took place in the 1950s, 1970s and 2000s. In his analysis, the author underlines that the Sámi manoeuvred the administrative setting with success and were able to use it as a resource in alignment with the planning officials.

In Chapter 8, Heta Lähdesmäki investigates changing forests and the perceived 'proper' place for wolves in twentieth century Finland. The author ventures out to investigate the case of Western Finland, where wolf packs are now observed after a long period of absence. According to Lähdesmäki, local people have argued that wolves should not live in Western Finland because there are no wilderness areas there. This reflects a perceptual frame in which many Finns see the wolf as a wilderness species. The author investigates the origin of this perception, while looking into wolves' history in the Finnish context.

Mauri Soikkanen and Simo Laakkonen argue that environmental history is a new way to explore World War Two, and focus on the mobilisation of natural resources in the Finnish context. They seek to uncover the role of hunting and fishing during WWII. To achieve their objective, the authors examine hunting and fishing along the thousand-kilometre border between Finland and the Soviet Union and focus on the largest wilderness area in Europe where Finnish soldiers found an abundance of wildlife species, which was dramatically affected by ensuing armed conflicts in the area.

Section 3 collects contributions that focus on the human alteration of the environment. In Chapter 10, Janne Mäkiranta investigates urban air quality in Finland from the late nineteenth century to the late 1960s. This essay shows how nineteenth and early twentieth century concerns over urban air were based on the enduring thought that clean air was beneficial for health. During the twentieth century, this vague hygienic idea was replaced by a more specific view deriving from medical research in industrial environments. This transition was fostered by public health officials who saw scientific knowledge as a way to deal with increasing indignation over urban air quality in Finland. From the late 1950s onwards, air pollution measurements and medical research provided a detailed analysis of air quality. As a result, the concern about urban air was directed towards specific pollutants and their potential effects on health, and was embraced by the environmental critics of the late 1960s.

In the last chapter, Tuomas Räsänen examines the relationship between humans and wildlife in Finland during the long twentieth century. The author identifies three gradual changes in the human-nature relationship. According to Räsänen, the first such turning point arose in the late nineteenth century, when the government monopolised and institutionalised the control of wildlife and launched extensive campaigns to wipe out unwanted species, such as large carnivores and predatory birds. Dissenting voices that advocated the protection of at least some persecuted species grew louder during the first decades of the twentieth century. This new attitude culminated in 1923 with the enactment of the first law for nature conservation, in which several critically endangered species were protected. The continued killing of endangered species led to the third turning point in the 1960s and the 1970s, when Finns gradually began to cherish and celebrate wild places and the wildlife that occupied them. This turnaround was displayed in a completely new way of caring for and nurturing animals. However, the care for animals has been reserved for select species, while the rest have been left aside, so animal populations in some cases bounced back from their low points, but several animal species continued to decline.

In *Seeing Like A State*, James C. Scott's influential book about the intersections of the natural environment and state intervention, Scott asserts that bureaucratic regimes aim to organise their societies according to the technocratic principles of 'high modernism', thus dooming their projects to failure and their societies to oppression.[9] For Scott 'large scale capitalism is just as much an agency of homogenization, uniformity, grids, and heroic simplification as

9. Scott 1998.

the state, with the difference that, for capitalists, simplification must pay'.[10] If this is so, it should not be surprising that Finland, despite its excellent record as a 'green superpower', has frequently prioritised exploitative practices over environmental concerns to achieve much desired economic prosperity. In the essays that follow, some commonalities emerge that unite environmental stories of Finland. First of all, contributions will reflect on accelerated ecological and societal changes, especially those of the last six or seven decades, which drastically remodelled human-nature relations in and around Finland. Second, most of the essays will juxtapose the duality of the cultural construction of the environment and the natural environment itself, which leads us to the third point – how the domestic social perception and the 'exported' images of nature in Finland seem to inhabit a diverging dichotomy from a grim reality, particularly in the post-1945 era.

This raises an interesting question: with a controversial and disturbing environmental history such as the ones reflected upon in the essays in this volume, can it be maintained that Finland has been an environmental success story, in other words a 'green superpower'? Or perhaps, by looking behind the façade of greenwashing and social hypocrisy, there is a more layered and complex image to be found. It is the hope of the editors that the questions raised by the volume will inspire further research and eventually contribute to a less exploitative human-nature relations in the Nordic countries and globally, as well as more open and realistic societal discourses about pressing ecological problems in the Finnish backyard.

10. James C. Scott, 'The trouble with the view from above', *Cato Unbound: A Journal of Debate*: https://www.cato-unbound.org/2010/09/08/james-c-scott/trouble-view-above (accessed 15 March 2018).

Bibliography

Anker, P. 2020. *The Power of the Periphery.* Cambridge: Cambridge University Press.

Friends of the Earth Finland. On UPM-Uruguay mega investment and the responsibility of Finland to secure and monitor that it respects economic, social and cultural human rights in compliance with the ICESCR obligations (2017): https://maanystavat.fi/sites/default/files/attachments/upm-uruguay_mega_investment_and_responsibility_of_finland.pdf (Accessed 5 Oct. 2022).

Hasler, B., K. Hyytiäinen, J.C. Refsgaard, J.C.R. Smart and K. Tonderski. 2019. 'Sustainable ecosystem governance under changing climate and land use: An introduction'. *Ambio* **48**: 1235–1239.

Jalava, J., R. Asplund, J. Ojala and J. Eloranta. 2006. *The Road to Prosperity: An Economic History of Finland.* Helsinki: Suomalaisen Kirjallisuuden Seura.

Jarridge, F. and T. Le Roux. 2021. *The Contamination of of the Earth.* Cambridge, MA: MIT Press.

Kuisma, M. 2006. *Metsäteollisuuden maa: Suomi, metsät ja kansainvälinen järjestelmä 1620–1920.* Helsinki: Suomalaisen Kirjallisuuden Seura.

Lettenmeier, M., C. Liedtke and H. Roh. 2014. 'Eight tons of material footprint – suggestion for a resource cap for jousehold consumption in Finland'. *Resources* **3**: 488–515.

McNeill, J.R. 2001. *Something New Under the Sun.* New York: W.W. Norton & Company.

Scott, James C. *Seeing Like a State.* New Haven: Yale University Press, 1998.

Winiwarter, V. et. al. 2004. 'Environmental history in Europe from 1994 to 2004: Enthusiasm and consolidation'. *Environment and History* **10** (4): 501–30.

SECTION 1

IDEAS AND THE HUMAN CONSTRUCTION OF THE ENVIRONMENT

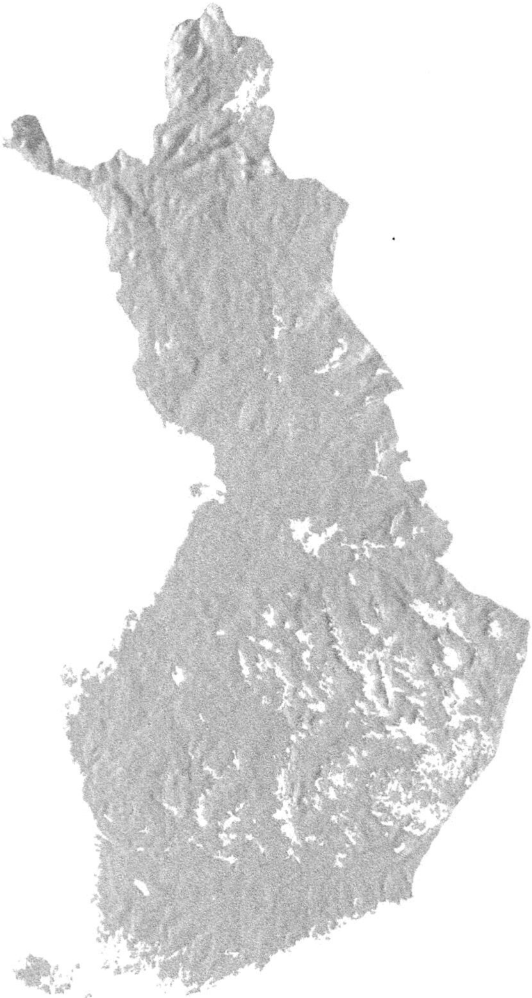

Chapter 2

KNOWLEDGE OF TREES AND FORESTS – FINNISH FOREST RESEARCH FROM THE NINETEENTH TO THE TWENTIETH CENTURY

Jaana Laine

Introduction

Forests are an essential part of Finnish culture and society in history and in the present and will be so in the future.

Research-based knowledge on forests has been and is integral both in societal discussion and the sustainable use of forests. This article explains how forest research over the past hundred years has gained and offered knowledge on trees and forests for state administration, societal stakeholders and private citizens. The article scrutinises the 'dialogue' between society and forests: what demands society has set on forests and, more importantly, what knowledge society has needed when living in coexistence with forests. The article sets forest research in the context of forest history and environmental history.

Finnish forests are mainly boreal forests with an abundance of coniferous trees. Around 79 per cent of the forests are dominated by Norway spruce (*Picea abies*) and Scots pine (*Pinus sylvestris*). The third dominating tree is the silver birch (*Betula pendula*). Forests cover around 75 per cent of the land area, which demonstrates forests to be the country's most important natural resource.[1] Traditionally, forests have been used in almost all aspects of everyday life: food, buildings, lighting and heating. Forests have offered raw materials for domestic use as well as for all forms of the forest industry.

1. Luke, Statistics database, Forest resources: https://statdb.luke.fi/PXWeb/pxweb/en/LUKE/.

doi: 10.3197/63824846758018.ch02

Up to the 1990s, forests were the basis of economic growth and prosperity. Until the mid-1950s, the forestry and wood and paper industry accounted on average for one fifth of GDP in Finland. Due to the diversification of the national production structure, the forestry sector's share of GDP has steadily decreased, and in the 2000s it has been below five per cent. Another important aspect is the forest industry's role in export revenues. Up to the 1950s, forest products generated an incredible eighty per cent of export revenues, declining below fifty per cent in the 1970s to around twenty per cent in the 2010s. In view of the above, the environmental history of forests is tightly embedded in Finnish economic and social history and needs to be contextualised accordingly.[2]

Because of the importance and multi-use of forests, through the centuries the main question has been how to balance sustainable and destructive uses of forests. In a society that has been and is economically, socially and culturally intertwined with trees and forests, knowledge of them has always been highly appreciated. Forest research has, from the late nineteenth century onwards, produced knowledge that societal actors have demanded, and even more important knowledge which they had not thought to ask for. In forest research, due to the slow growth of trees, the timespan of creating new knowledge is often counted in decades. Forest research stresses careful planning and foresight about the expected demands of knowledge.

This article includes four main sections, covering different periods: Know the Forests (late nineteenth century–1930s), Exploit the Forests (1940s–1960s), Define the Forests (1970s–1990s) and Discover Forests' Futures (2000s–). Each section consists of a short description of societal circumstances, clarifying the status of and demands placed on forests during that specific period, and paragraphs explaining the specific forest-related knowledge of the period.

Know the forests (late nineteenth century–1930s)

Through the nineteenth century and up to 1917, the Grand Duchy of Finland found herself under the rule of Russian tsars. However, the forest-related institutions and scarce legislation reflected mostly Swedish origins, despite Swedish rule having ended in 1809. The export of tar and sawn timber formed the major industrial use of forests. Forests played an elemental role in household consumption (housing, fuel, nutrition, etc.) and it was not until the beginning of the 1920s that industrial use of timber exceeded household consumption.[3]

2. Laine 2019.
3. Kunnas 1973, pp. 110–11.

The reform of forest legislation from the mid-nineteenth century onwards, and especially the 1857 repeal of production restrictions on steam-powered sawmills, enhanced the economic use of forests. When pulp and paper mills, mainly from the 1880s onwards, joined this development, the wooden path towards prosperity was mapped out.[4] The most far-reaching institutional changes took place after independence in 1917 and the civil war in 1918, when land reform, the so-called Crofter Liberation, transformed most former leaseholders into landowners. Widespread private forest ownership emerged, and forests were defined as part of farms.

Forests being the most valuable natural resource in Finland, the state demanded more information both about existing stock volume and its quality. This task was entrusted to German forester Edmund von Berg, who after travelling mainly in Southern and Central Finland, wrote his report in 1858. In this first evaluation of the Finnish forests, von Berg stated that the situation was alarming, especially around towns and major traffic routes, mainly due to cutting of fuel wood and traditional slash-and-burn cultivation.[5] The following decades, witnessing ever-growing forest industrial production, increased stakeholders' uncertainty about forest resources and the need for university-educated foresters and science-based forest research.

Young botanist A.K. Cajander was persuaded to write a proposal for the organisation of forest research, which he did in 1906, after visiting several German and other European forest research organisations. Due to the second oppressive Russification period (1909–1917), the First World War (1914–1918) and the Finnish civil war (1918), the forest research organisation only began to operate at last in 1918. Forest research and education formed a tripartite structure, where the Finnish Forest Research Institute co-operated with the University of Helsinki, responsible for higher forest education, and the Society of Forest Science, which shared research results and information with the stakeholders in the forest sector.[6]

The first years of the Finnish Forest Research Institute were filled with ambitious expectations in extremely modest circumstances. The staff, consisting of three professors, represented the research fields of silviculture, forest mensuration and forest soil science. Later, four more disciplines entered the

4. Michelsen 1995.
5. Berg [1859] 1988.
6. Laine 2017, pp. 22–24.

research institute: in 1928 peatland forestry and in 1938 forest economics, technology and biology.[7]

Forest regeneration and management

According to research publications in the 1920s, the forests seemed to be in a rather poor condition. Due to unclear ownership institutions, high household consumption, active tar production and steadily growing industrial use of timber the destruction of forests was feared. Therefore, one major task for the newly founded research institute was to develop silvicultural practices, e.g., regeneration, thinnings and cuttings. They concentrated on forest regeneration, both natural and artificial. Natural regeneration in northern Finland was carefully scrutinised since the harsh northern climate limited seed production in these areas, where state-owned forests were largely situated. Researchers concentrated on different phases of natural regeneration, such as soil preparation with a horse-driven harrow. Even though natural regeneration covered nearly all forests, the first steps towards artificial regeneration, via nurseries, domestic and foreign seed provenances and forest tree breeding, were also taken.[8]

Towards the end of the 1930s, researchers focused on forest management practices. Traditional selective cutting, high-grading where mainly older and bigger trees were cut, was increasingly deemed unsatisfactory among forest professionals. Scientific research and public discussion anticipated the changes in forest management towards clear cutting and period cover forestry that were enacted after the Second World War.

National forest inventories

The questions set by societal stakeholders and state leaders were, how many trees, of what kinds, and where. Until the beginning of the 1920s, all information about forests was vague estimation lacking a scientific basis. The first national inventory of forests, conducted in 1921–1924, opened a long-lasting series of inventories of which the latest was the thirteenth in 2021.[9] The first and the three following inventories (1936–1938, 1951–1953 and 1960–1963) involved symbolic and heroic achievements. Skilful forest professionals walking through

---

7. Ibid., p. 24.

8. Heikinheimo 1939.

9. Forest resources, www.luke.fi.

the country measured and evaluated forests. For instance, during the second inventory, the total length of inventory lines was around 24,600 kilometres.[10]

Besides current information on standing stock, they also estimated the future need for forest management activities and the possible drain on standing stock. For the forest industry, the information was published according to geographical areas and watercourses, since industrial timber procurement was highly dependent on log floating.

Information gathering was not restricted merely to growing trees –other flora was also reported with exceptional accuracy. The main reason for this lies in the prevailing paradigm for Finnish forest management and increment, the forest type theory created by Professor A.K. Cajander. The most suitable and profitable tree species for the forest concerned are extrapolated based on the forest ground flora.[11] This paradigm still dominates Finnish forestry in the 2020s.

Timber consumption inventories

To ensure national prosperity, information on the amount of growing stock and forest productivity was not enough. Nearly as important was information on timber consumption. For what purposes was the cut timber used, and how did this contribute to economic development? Accordingly, the Forest Research Institute launched the first national timber consumption inventory in 1927. Information was gathered all over the country, from forest industries, municipalities, forest-related organisations and even from 1,400 private farms that reported all their timber consumption for a year.

The inventory revealed astonishing details about the wood-dependency of Finnish society in 1927. The statistics consisted of amounts of timber used in pulp mills, sawmills or transportation (e.g., steamships and trains) and export of roundwood. Industry used 49 per cent and export of roundwood accounted for eleven per cent of total wood consumption (40.14 million m^3). However, from the historical research point of view, the statistics about wood used for heating are also fascinating, e.g. heating of schools and hospitals, or how often Finns heated the sauna in summer compared with winter. The original questionnaires included information on the size of houses, the number of fireplaces, the number of inhabitants and livestock. Domestic use formed 32 per cent of all wood consumption and two-thirds of this was for heating.[12]

10. Laine 2017, pp. 33–35, 51–53.
11. Cajander 1949.
12. Saari et al. 1934.

In 1929, statistics stated that the average size of the 232,400 privately owned forests was 45 hectares and 75 per cent of all private forests were under fifty hectares.[13] This raised concern in the forest industry, since timber procurement relied mostly on private forests – around two thirds of the timber used in the forest industry originated from private forests.[14] Small forest areas, firstly, diminished forest owners' ability and willingness to sell timber and, secondly, raised costs the timber procurement.

The first decades of forest research stressed the need to know the forests. Researchers produced knowledge on forest resources and wood consumption, besides new silvicultural practices.

Exploit the forests (1940s–1960s)

The two decades after the Second World War are characterised by intensive development of forest institutions, management and practices. Forest research concentrated on the improvement of forest management and technology, the regeneration of low-yielding stands, the ditching of peatlands and gaining more information on forest work and workers.[15]

For Finland, the Second World War ended in a catastrophe. Finland was condemned to pay US$300 million (at 1938 prices) war reparations to the Soviet Union. Territorial concessions, amounting to ten per cent of land area, caused the loss of twelve per cent of forests and twenty per cent of industrial capacity, consisting mainly of the forest industry.[16] War reparations concentrated heavily on the products of the steel and engineering industry, which helped to diversify the structure of the national economy. However, both export revenues and citizens' income were still highly dependent on forests and the forest industry.

Although the war caused nearly 100,000 Finnish casualties, the civilian population mostly remained untouched. However, the people of ceded areas, around 420,000 evacuated citizens (circa eleven per cent of the total population), had to be resettled. Around half of them continued as farmers and forest owners through the resettlement process, and nearly 52,000 new farms reclaimed land mainly from municipalities, state and companies, but also from private citizens. Post-war resettlement slowed down the modernisation of Finnish society and,

---

13. Osara 1936.
14. Laine 2006, pp. 93–96.
15. Uusitalo 1978.
16. Eloranta et al. 2006.

at the beginning of the 1950s, nearly half the workforce was still occupied in agriculture and forestry.[17]

The post-war years stressed efficiency and productivity. The administrative grip over private forest owners tightened. New and more specific regulations on silviculture and timber felling in private forests also included compulsory membership of local Forestry Management Associations. State administrative District Forestry Boards enforced the Private Forest Act and, if the law was not respected, private forest owners were subject to fines, the forest was taken under supervision and owners' rights to decide forest management were restricted for years. Forests and economic-based forestry became a symbol of national consensus, as demonstrated in June 1950 by the Forest March. Over six summer days, around half a million Finns, eighteen per cent of the population over fifteen years of age, participated in silvicultural activities.[18]

After World War Two, the state research organisations developed extensively. Forest research received its share of resources and established a nationwide network of research stations. The most radical change, however, occurred within international research co-operation and the language used. Forest science, as well as other sciences, had been in many aspects connected to the German forest tradition. Suddenly, after the war, forest science and researchers turned to North America. It was a confusing situation. Highly respected forest scientists had published their research in German, and most barely spoke English. This transition offered opportunities for younger scholars and, until the late 1960s, new professors stepped forward.[19]

After somewhat reckless wartime cuttings, silviculture and forest management needed to be reorganised. The selective felling of large trees, quite common before the war, was banned and private forest owners were forced to adopt period cover forestry with clear cutting and artificial regeneration. Despite more efficient timber procurement, the lack of timber seemed to restrict the forest industry's capacity to grow. To alleviate the timber shortage, massive forest improvement activities were launched.

Renewed forest management

During the 1940s, a fundamental paradigm shift took place in Finnish forestry. The signs of change already existed in the late 1930s, but the real starting point

17. Ilvessalo 1963.
18. SVT 6:C 102,8, Väestötilastoa Population Census 1950, p. 18.
19. Laine 2017, pp. 58–63.

was the Declaration against selection felling in 1948. With this declaration, six distinguished foresters rejected selection felling and proclaimed the benefits of silvicultural fellings, i.e., clear cuttings and artificial regeneration with saplings. Forest professionals, supported by the state administration, implemented this change in practice and legislation.[20]

Forest researchers faced the challenge of providing coherent information to guide thinning, clear-cutting and regeneration. They created specific tables for different geographical areas, forest types and tree species. Based on these tables, forest professionals decided the timing of thinning and other cuttings. When implementing the tables, the volume of thinning was based on the defined basal area (with a specific gauge) and estimation of the average height of trees. This method systematised and standardised silviculture all over the country and aimed to secure timber for industrial use.[21]

The forest industry's demand for timber grew inexorably and finally threatened foresters' main principle – that yearly cuttings should not exceed the yearly increment. However, large clear cuttings with evolving harvesting technology had caused a situation where the volume of growing stock seemed to decrease. Something needed to be changed. The solution was to improve the increment of the growing stock by launching massive forest improvement campaigns.[22]

Forest improvement

From the 1960s onwards, the state launched several national forestry improvement programmes, funded by the Bank of Finland and partly by the World Bank. These programmes estimated the connection between total timber consumption, growing stock and increment, maximum removals and, especially, how and with what measures to increase timber volume. The major funded activities were forest tree breeding, artificial regeneration, afforestation of low yielding forests and abandoned fields, drainage and fertilisation. In addition, the state funded building of forest roads and development of forest work technologies. Since, according to the fifth national forest inventory (1964–1970), private citizens owned 58 per cent of forestry land and 71 per cent of growing stock, funding was especially awarded to the privately owned forests. Private forest

20. Laine 2017, pp. 54–55.
21. Nyyssönen 1954.
22. Uusitalo 1976, pp. 105–35.

owners commonly paid half of the forest improvement costs themselves and the state contribution was 21 per cent loan and 29 per cent subsidy.[23]

Finland is a swampy country. One-third of the land area (9.2 million hectares), classified as various swamps, bogs or peatlands, was evaluated as a solution to the looming timber shortage. National planning supported by specific funding and administration officers worked efficiently. At the beginning of the 1950s, drainage covered around eighteen per cent of the swamps in southern Finland and four per cent in northern Finland. In the next decade, the shares were 47 per cent and 21 per cent, an incredible achievement given the modest technology available and hard conditions.[24] Unfortunately, overly effective and motivated work resulted in errors, which have since been widely criticised. Draining also affected around 0.45 million hectares of nutrient-deficient swamps and bogs, which lacked timber production capacity. In the twenty-first century, these areas have mainly been left outside active forestry or are under restoration back to swamps. Thus, peatland forests, representing around 23 per cent of growing stock and twenty per cent of increment and removals, have a growing importance in Finnish forestry.[25]

One highly respected but criticised measure of forest improvement was forest tree breeding. When entering plantation forestry, large clear cutting areas and low yielding forests were regenerated with saplings, and the need for inherited high-quality seeds was urgent. Northern Finland, where a good seed year might occur every tenth year, particularly suffered from lack of seed. Tree breeding is based on reproducing genetically inherited properties such as healthy trees with straight stems and better increment. Finnish Forest Tree Breeding and Forest Research Institute created a nationwide system of so-called plus trees whose seeds were collected and used in tree nurseries. Later, specific seed orchards fulfilled the need for seed and, in the twenty-first century, three-quarters of all saplings originate from seed orchards.[26] Despite good results, forest stakeholders have criticised tree breeding as an extremely expensive and slow way of increasing productivity.

23. Laine 2017, pp. 68–71.

24. Ibid., p. 56.

25. Luke, Suometsät: https://www.luke.fi/tietoa-luonnonvaroista/metsa/suometsat/.

26. Rusanen et al. 2021; Laine 2017, pp. 76–80; Luke, Tree breeding: https://www.luke.fi/en/natural-resources/forest/tree-breeding/.

Forest work

Forest work – timber felling, hauling and floating – was regarded as a basic skill for rural men. In 1950, nearly seventy per cent of rural men worked in the forest at least a few days a year. This common knowledge faded as Finnish society slowly recovered from the war and industrial employment lured rural young men away from the forest. A shortage of skilled forest workers was threatening timber supply and forest industry production.[27]

Rising societal prosperity contrasted to forest workers' hard conditions and modest income. Their work and life became a political question, which culminated in President Kekkonen's rhetorical claims that forest workers' poor living conditions were a national disgrace. Men whose hard work enabled Finnish society to gain prosperity lived in misery.

At first, forest research improved practices and tools for forest work and aided stakeholders, employers and trade unions in defining piece wages. Later, research was impacted by the social sciences and the focus was laid on forest workers' professionalism, societal position and living and health issues. Results and especially popularised publications raised a mix of embarrassment and guilt which also benefited forest workers.[28] At the beginning of the 1970s, they became qualified professionals with a new modern title and a better labour market position.[29]

Define the forests (1970s–1990s)

Finnish society finally became modern. Up to the 1950s, the majority of Finns were occupied in agriculture and forestry. During the 1960s, Finland pretty much jumped straight from being a primary sector society to a tertiary sector society. That is, Finland skipped the phase where citizens occupy themselves mainly with industrial tasks.[30] Exceptionally quick and somewhat violent changes in social structure caused severe problems: rural citizens moved to crowded cities struggling with the housing shortages whereas rural areas were deserted and small farms collapsed with diminishing agriculture. The turn of the 1960s and 1970s witnessed Finnish mass emigration to Sweden in search of a better life.[31]

27. Laine 2019.
28. Heikinheimo 1972.
29. Laine 2012.
30. Eloranta et al. 2006.
31. Koivukangas 2003.

Like other Western countries, Finland encountered two significant global phenomena in the 1970s. Firstly, the golden years of economic growth ended abruptly in oil crises, causing economic recession and unemployment. This economic shock was smoothed by the bilateral trade with the Soviet Union whereby imported oil was paid for by the export of Finnish products to the Soviet Union.[32] Secondly, the environmental movement also arrived in Finland, starting several decades of harsh disputes and conflicts about forests.

Forest research reacted quite actively to the oil crises, which increased interest in fuel wood. The forest industry had no intention of giving any timber for heating purposes; all timber was badly needed for industrial use. Therefore, research concentrated on substitutes, such as growing willow as the source of energy.

The image of forestry and the forest industry turned to that of a sunset business when, in the 1990s, Nokia and other electronics industries began their triumphal success. Towards the end of the 1990s, the forest sector and research had to justify the societal importance of their activities.

Environmental conflicts and criticism against forest management in the 1970s were hardly acknowledged in forest research, which was committed to economic-based forest management. However, urbanised citizens, enjoying forests more for recreational purposes, embraced environmental and conservation values. Little by little, forest research broke away from economy- and ecology-based research and new, multidisciplinary research with more societal themes arose. Forest research came to acknowledge the recreational and conservational importance of forests.[33]

Acid rain

Concern for the future of forests also shook Finns. News and photos of Central European forests dying because of air pollution and acid rain alarmed the forest sector and society. Could this be possible also in Finland? In the face of growing unrest, forest research discovered conifers suffering from needle loss, especially in northern Finland. Since these northern areas almost totally lack domestic industrial activities, the source of acid rain had to be outside the country's borders, especially in the Russian Kola Peninsula. Acid rain, the threat of dying trees and needle loss created strong debate both among the public and forest researchers. The official results of forest research declared that Finnish nature

32. Eloranta et al. 2006.

33. Laine 2017, pp. 101–13, 137–41.

did not significantly suffer from acid rain. The damage to conifers was caused mainly by natural phenomena, such as abundant rainfall and early autumn frost. However, public discussion and a suspicious attitude towards the forest sector and research had gained more room, normalising the questioning of the forest sector's authority and paving the way for continuous forest conflicts.[34]

The acid rain discussion had some positive aspects in terms of developing multidisciplinary research. In the mid-1980s, the state funded one of the first multidisciplinary research programmes to study the consequences of air pollution and acid rain. Altogether eight state research organisations, seven universities and around 200 researchers scrutinised, over five years, the amount of acidification and its impact on nature and forests. Next, another five-year research programme concentrated on how air pollution affected forests. All in all, the 1980s opened the period of multidisciplinary research and directed it towards taking into account not just forests but also other environmental and societal issues.[35]

Important economy

Forests are the source of national prosperity and enable the forest industry. Thus, effective technologies, procedures and institutions are the prerequisites for a prospering forest sector; economics has been an integral part of Finnish forest research from the late 1930s onwards.

The key theme of the economic research has been roundwood trade, its practices and especially its prices. From the 1960s onwards, the Central Union of Agricultural Producers, representing private forest owners, and the Central Association of Finnish Forest Industries agreed on the price of roundwood. These agreements set the prices for different timber assortments (logs and pulpwood of various tree species). At the beginning of the 1990s, due to EU regulation and the forest industry's motivation to withdraw from the cartel, the co-operation over price setting stopped. However, despite anti-trust legislation, forest industry companies still secretly agreed on the roundwood prices. This illegal cartel ended in 2004.[36]

Research on roundwood trade and published forecasts evaluating price trends often caused conflicts between forest researchers and the forest industry. Industry representatives even occasionally declared that forest researchers

34. Ibid., pp. 144–47.

35. Ibid., p. 145; Tikkanen and Niemelä 1995.

36. Kuuluvainen et al. 2021.

should do what they are capable of and leave important business (i.e. timber trade) issues in the hands of the forest industry. The Forest industry's cautious attitude was due to its dependence on private forests but also to the possibility that forecasts might have an unwanted impact on roundwood trade.[37]

Up to the 1980s, studies concentrated on private forests' importance for agriculture and forest owners' incomes. Recently, research on forest owners' values and attitudes and their willingness to cut trees have become more common. This research has remained important mainly due to the private forests' irreplaceable role in the forest industry's timber procurement. In the 2010s, the forest industry used around 55 million cubic metres of domestic roundwood, of which 75 per cent originated from privately owned forests. The latest research on private forest owners was published in the spring of 2020.[38]

Discover forests' futures (2000s–)

Economic recessions and recoveries have characterised Finnish society since the 1990s. It has searched for direction and solutions to achieve a balance between economic growth and environmental issues. The forest sector and industry, deemed outdated sunset businesses, have encountered new appreciation in the course of globally increased demand for, e.g., tissue paper. Both new technologies and wood-based products and a strengthening understanding of forests' role in mitigating climate change have altered the societal position of forest-based activities.

Societal changes and attitudes set challenges to forest management and knowledge. Especially on social media, environmental and forest sector stakeholders react fast – and sometimes even furiously – when they think that 'the others' threaten their forest-related values and aims. Far too often, reactions and responses cause disputes and even escalate conflicts. Future forests require active reconciliation of different societal and global objectives.

Especially in the face of climate change, forests have increasingly gained global significance alongside national ones. In addition to national objectives and institutions, Finnish forests are subject to the European Union and other international obligations. Forest research aims to effectively develop co-operation and understand the complex future of forests.

Recognising the new possibilities in forests and the forest sector, society demanded fundamental changes in forest research. This needed the creation of

37. Laine 2017, pp. 193–94.

38. Karppinen et al. 2020.

a more multidisciplinary approach to forest research, particularly in the social and natural sciences. In addition to economic values, other approaches to sustainability (ecological, societal and cultural) had to be embedded in the research projects. Two main research themes, climate change and the bioeconomy with new products, have influenced almost all research.

Although international co-operation between forest research organisations has been taking place within IUFRO for a hundred years, the nature of the co-operation is expanding and becoming more multidisciplinary. Solving major global problems requires strengthening scientific co-operation within the European Union and other international organisations.

New research trends exist in forestry research. Firstly, almost all studies notice climate change in one way or another and seek to find out how ongoing research could mitigate climate change. Secondly, researchers are working to develop new forest-based products and promote bioeconomy. Thirdly, we find the somewhat revolutionary ideology of post-humanism, which aims to understand the forest as an entity where values are not defined only from a human perspective.

Recreation and wellbeing

In Finnish legislation everyman's right, or freedom, to roam lays all forests open to citizens despite their ownership. Everyone has the right to enter the forest and enjoy its benefits. Simultaneously, as the wellbeing of civil society has developed, citizens' everyday life has become detached from the forest. Many citizens no longer value the forest as a place of work or source of income. Forests are more valued for their recreational purposes and other than economic values.[39]

Research on forests' recreational importance has developed from being undervalued towards being an essential part of forest research's societal impact. The national outdoor recreation inventory, the first enacted in the late 1990s and the third by the end of 2021, has offered knowledge on citizens' attitudes toward nature, environmental changes and recreational values. According to the second inventory (2009–2010), around 96 per cent of citizens engage in outdoor activities two or three times a week. Besides recreation, the health impact of forests and nature-based tourism have opened new business opportunities for rural areas.[40] In 2020–2022, due to COVID-19, national parks experienced an exceptional rush. The surveys, conducted in co-operation with the Finnish

39. Simkin et al. 2021.

40. Sievänen et al. 2011.

state forest organisation and other Nordic countries, have explained both the number of visitors in national parks and their values and intentions. In 2011, researchers counted more than two million visits to Finnish national parks, in 2015 there were around 2.6 million visits, and finally 2020 saw around four million visits.[41] For a population of 5.5 million, these numbers are significant. The pandemic pushed Finns back to nature and forests, which will presumably have an impact on forest-related values. An impression based on public discussion sees ecological values outweighing economic ones.

The popularity of the berry crop forecasts is explained by the universal right to pick berries and mushrooms despite forest ownership. Every second Finn picks berries or mushrooms. The most important wild berries are bilberry and lingonberry, and cloudberry in the North. It is estimated that the yearly amount picked of wild berries is 40–45 million kilos, around ten per cent of the total crop.[42] Mushrooms are the other most often used product of forests and the amount of mushrooms picked varies yearly between three and sixteen million kilos. Researchers prepare separate mushroom yield forecasts for enthusiastic mushroom pickers.[43]

Climate change

Climate change has altered the way we look at our forests. For centuries, forests have provided us with game, berries, mushrooms and wood. Traditionally, forests have also protected against enemies, in the Finnish case neighbours or soldiers of the Swedish king or the Russian tsar. Our everyday life is based on forests and our culture has drawn its inspiration from it. Forests, important for carbon sequestration, have gained new value as carbon sinks and storage. Most standards and activities within forest management need revaluation.

Forests and forest management are facing several new difficulties. A warming climate is encouraging the emergence of new pests such as bark beetles, which can cause the widespread deaths of spruce trees. Increased rainfall and storms will have a negative impact on soil and trees. Preparing for climate change requires altering forest management practices, perhaps new tree species being used in regeneration and ensuring that forests after felling can withstand increased winds and storms. An effort is made on peatland forests, as drainage

41. Laine 2017, p. 231; Metsähallitus, Käyntimäärät maastossa: https://www.metsa.fi/vapaa-aika-luonnossa/kayntimaarat/kayntimaarat-maastossa/.

42. Roininen and Mokkila 2017, p. 12.

43. Luke, Mushroom yield: https://www.luke.fi/en/natural-resources/forest-berries-and-mushrooms/mushroom-yield/.

turns these areas into a source of carbon dioxide. Therefore, logging peatland forests demands special consideration.[44]

Bioeconomy and novel products

The raw materials from forests are seen as an increasingly diverse resource, allowing forest industry production to evolve towards the bioeconomy and even the circular economy. One highly fascinating research field focuses on bioactive compounds extracted from forest-derived materials. Current research produces new discoveries and verifies old traditional knowledge, for instance, the anti-inflammatory effect of resin. The new compounds are used not only in the pharmaceutical industry but also in the nutrition and cosmetics industries.[45]

Forests as entity

Alienation from the economic use of forests and forest work, as well as urbanisation, have changed the values attached to forests. Increasing numbers of people appreciate the non-economic importance of forests and see the forest as an entity, not just as a timber resource. Forests valued as recreational sites should remain intact, preferably with old-growth trees, rich biodiversity and efficient carbon sink and storage. Occasionally, economic-oriented forest owners and forestry professionals have difficulty in understanding non-economic aims, not to mention post-humanistic attitudes.

Conclusion

The knowledge needed and gathered from the forest has varied through the decades. The beginning of the twentieth century witnessed an ever-growing need for knowledge on the volume and quality of growing stock. The post-war era concentrated on forest improvement and searched for practices to increase timber growth and develop forest management and timber procurement. Due to the environmental movement, the focus widened to the non-economic aspects of forests, such as recreational and health issues. Finally, in the twenty-first century, some sectors of forest research scrutinise forests as an entity without assessing only economic significance. Forests, with their different living organisms and non-living elements, are valuable.

---

44. Venäläinen et al. 2020.

45. Högbom et al. 2022.

The main currents of forest research include, firstly, forest management-related knowledge and activities; secondly, research discovering the relationship between society and forests; thirdly, research on rich biodiversity; and finally, understanding the forest as an ecological entity in its own right.

Throughout the century, forest research has responded to the information needs of Finnish society, either by providing forward-looking accounts or by assessing developments that have already taken place. Either way, knowledge has shaped societal decision-making and strengthened forests' impact on the development of society.

Bibliography

Berg, Edmund von. [1859] 1988. *Kertomus Suomenmaan metsistä*. Helsingin yliopiston metsänhoitotieteen laitoksen tiedonantoja 63. Helsinki: Helsingin yliopisto.

Cajander, A.K. 1949. 'Forest types and their significance'. *Acta Forestalia Fennica* **56** (5): 1–71. Helsinki: Finnish Society of Forest Sciences. http://hdl.handle.net/10138/17982

Forest resources. Luonnonvarakeskus: www.luke.fi (accessed 15 Sept. 2021).

Eloranta, Jari, Jalava Jukka and Ojala Jari (eds). 2006. *The Road to Prosperity: An Economic History of Finland*. Helsinki: Finnish Literature Society.

Forest Europe. 2020. State of Europe's Forests 2020: https://foresteurope.org/wp-content/uploads/2016/08/SoEF_2020.pdf

Heikinheimo, Lauri. 1972. *Suomalainen metsätyömies*. Porvoo: WSOY.

Heikinheimo, Olli. 1939. 'The Forest Research Institute of Finland and its activities from 1918 to 1938'. *Communicationes Instituti Forestalis Fenniae* **28** (1): 1–39. Helsinki: Forest Research Institute in Finland. http://urn.fi/URN:NBN:fi-metla-201207171060

Högbom, Lars, Jõgiste Kalev, Kniivilä Matleena, Lukmine Diana, Mustonen Mika, Rautio Pasi, Samariks Valters, Svensson Johan, Vodde Floortje, Zute Daiga and Øistad Knut. 2022. Future trends in forest management. The Profor – Policy Brief. Helsinki: Natural Resources Institute Finland. http://urn.fi/URN:ISBN:978-952-380-452-4

Ilvessalo, M. 1963. 'Suomen vuosien 1939–40 ja 1941–44 sotien jälkeinen asutustoiminta metsätalouden kannalta'. [Deutsches Referat: Die finnische Siedlungstätigkeit nach den Kriegen 1939–40 und 1941–44 vom Standpunkt der Forstwirtschaft]. *Communicationes Instituti Forestalis Fenniae* **56** (4): 1–184. Helsinki: Forest Research Institute in Finland. http://urn.fi/URN:§NBN:fi-metla-201207171088

Karppinen, Heimo, Harri Hänninen and Paula Horne. 2020. *Suomalainen metsänomistaja 2020*. Helsinki: Luonnonvarakeskus. http://urn.fi/URN:ISBN:978-952-326-961-3

Koivukangas, O. 2003. Finns abroad. A short history of Finnish emigration. Turku: Institute of Migration: https://arkisto.org/wp-content/uploads/2017/08/027_Koivukangas.pdf

Kunnas, Heikki J. 1973. Metsätaloustuotanto Suomessa 1860–1965 [Forestry in Finland, 1860–1965]. Suomen Pankin julkaisuja. Kasvututkimuksia 4. Helsinki: Suomen Pankki [Bank of Finland.] http://urn.fi/URN:NBN:fi:bof-201608301392

Kuuluvainen, Jari, Jaana Korhonen, Lanhui Wang Lanhui and Anne Toppinen. 2021. 'Wood market cartel in Finland 1997–2004: Analyzing price effects using the indicator approach'. *Forest Policy and Economics* **124** (102380): https://doi.org/10.1016/j.forpol.2020.102380

Laine, Jaana. 2006. *Puukaupan säännöt. Yksityismetsänomistajien ja metsäteollisuuden puukauppa Itä-Suomessa 1919–1939*. Helsinki: Suomen Tiedeseura. http://hdl.handle.net/10224/4081

Laine, Jaana. 2012. 'The journey of Finnish forest workers from woods to negotiating table: The first Collective Labour Agreement in 1962'. *Scandinavian Journal of History* **37** (3): 377–400. https://doi.org/10.1080/03468755.2012.671625

Laine, Jaana. 2017. *Metsästä yhteiskuntaan. Metsäntutkimuslaitos 1917–2012*. Helsinki: Luonnonvarakeskus and Metsäkustannus. http://urn.fi/URN:ISBN:978-952-338-102-5

Laine, Jaana. 2019. 'Metsä talouden ja arvojen risteyksessä'. In J. Laine, S. Fellman, M. Hannikainen and J. Ojala (eds), *Vaurastumisen vuodet. Suomen taloushistoria teollistumisen jälkeen.* Helsinki: Gaudeamus. pp. 131–49. https://helka.helsinki.fi/permalink/358UOH_INST/1b30nf2/alma9932250183506253

Luke, Mushroom yield: https://www.luke.fi/en/natural-resources/forest-berries-and-mushrooms/mushroom-yield/

Luke, Statistics database: https://statdb.luke.fi/PXWeb/pxweb/en/LUKE/.

Luke, Suometsät: https://www.luke.fi/tietoa-luonnonvaroista/metsa/suometsat/.

Luke, Tree breeding: https://www.luke.fi/en/natural-resources/forest/tree-breeding/.

Michelsen, Karl-Erik. 1995. *History of Forest Research in Finland. Part 1. The Unknown Forest.* Helsinki: Forest Research Institute in Finland. http://urn.fi/URN:ISBN:951-40-1471-5

Nyyssönen, Aarne. 1954. 'Metsikön kuutiomäärän arvioiminen relaskoopin avulla'. [Estimation of stand volume by means of the relascope]. *Communicationes Instituti Forestalis Fenniae* 44 (6): 1–31. Helsinki: Metsätieteellinen tutkimuslaitos. http://urn.fi/URN:NBN:fi-metla-201207171076

Osara, N.A. 1936. *Metsälötilasto vuodelta 1929.* Metsätieteellisen tutkimuslaitoksen julkaisuja 21. Helsinki: Metsätieteellinen tutkimuslaitos. http://urn.fi/URN:NBN:fi-metla-201207171053

Roininen, K. and M. Mokkila. 2017. *Selvitys marjojen ja marjasivuvirtojen hyödyntämispotentiaalista Suomessa.* Helsinki: Sitra. https://media.sitra.fi/2017/02/27173257/VTTn20marjaselvitys20b-2.pdf

Rusanen, Mari, Egbert Beuker, Leena Yrjänä, Matti Haapanen and Sanna Paanukoski. 2021. *Finland's Forest Genetic Resources, Use and Conservation. Natural Resources and Bioeconomy Studies 4.* Helsinki: Natural Resources Institute Finland. http://urn.fi/URN:ISBN:978-952-380-147-9

Saari, Eino, Paavo Aro, Eino Hartikainen and Viljo Pöntynen. 1934. *Puun käyttö Suomessa.* Metsätieteellisen tutkimuslaitoksen julkaisu 14, 1. Helsinki: Metsätieteellinen tutkimuslaitos. http://urn.fi/URN:NBN:fi-metla-201207171047

Sievänen, Tuija, Marjo Neuvonen and Eija Pouta. 2011. 'National Park visitor segments and their interest in rural tourism services and intention to revisit'. *Scandinavian Journal of Hospitality and Tourism* 11: 54–73. https://doi.org/10.1080/15022250.2011.638210

Simkin, Jenni, Ann Ojala and Liisa Tyrväinen. 2021. 'The perceived restorativeness of differently managed forests and its association with forest qualities and individual variables: A field experiment'. *International Journal of Environmental Research and Public Health* **18** (2): 422. https://doi.org/10.3390/ijerph18020422

Tikkanen, Eero and Irja Niemelä (eds). 1995. *Kola Peninsula Pollutants and Forest Ecosystems in Lapland: Final Report of the Lapland Forest Damage Project.* Helsinki: Forest Research Institute in Finland. http://urn.fi/URN:ISBN:951-40-1455-3

Uusitalo, Matti (ed.). 1976. 'Metsätilastollinen vuosikirja 1974. [Yearbook of Forest Statistics]'. *Folia Forestalia* 255. Helsinki: Metsäntutkimuslaitos. http://urn.fi/URN:ISBN:951-40-0200-8

Uusitalo, Matti. 1978. 'Suomen metsätalous MERA-ohjelmakaudella 1965–75: Tilastoihin perustuva tarkastelu. [Finnish forestry during the MERA programme period 1965–75: A review based on statistics]'. *Folia Forestalia* 367. Helsinki: Metsäntutkimuslaitos. http://urn.fi/URN:ISBN:951-40-0359-4

Venäläinen, Ari, Ilari Lehtonen, Mikko Laapas, Kimmo Ruosteenoja, Olli-Pekka Tikkanen, Heli Viiri Heli, Veli-Pekka Ikonen and Heli Peltola Heli. 2020. 'Climate change induces multiple risks to boreal forests and forestry in Finland: A literature review'. *Global Change Biology* **26** (8). https://doi.org/10.1111/gcb.15183

Chapter 3

'REACHING MATURITY' OR 'SELLING OUT'? THE IDEA OF GREEN GROWTH IN FINNISH GREEN PARTY ENVIRONMENTAL DISCOURSES 1988–1995

Risto-Matti Matero

Introduction

Over the past decades, major shifts have taken place in public environmental discourses transnationally. At the waking of the global environmentalist movement at the turn of the 1970s, new eco-philosophies emerged, from deep ecology to ecofeminism, questioning, among other things, the deep-rooted connection between wellbeing and growth economics. Environmental grassroots movements, in many ways the successors of 1960s radicalism, aimed to create a new culture separate from growing consumption and market dependence.[1] Starting from the early 1980s, Green parties were formed throughout Europe to represent these alternative social movements. Simultaneously, a more moderate form of environmentalism revolving around the idea of sustainable growth started developing.[2] By the turn of the millennium, a large chunk of earlier green radicalism had leaned into moderate ideals of greener growth and consumption, more easily adaptable to a free market political system. By the end of the 1990s, the ideals of green growth had bypassed earlier radical attempts at restructuring modernity and growth economics in public environmental discourses and in Green parties throughout Europe.[3]

1. Guha 2000, Hockenos 2008 and Milder 2019, for example, analyse the development of environmentalist movements.
2. Warde et al. 2018, pp. 69–70; Caradonna 2019.
3. Examined thoroughly in Borowy and Schmeltzer 2019.

doi: 10.3197/63824846758018.ch03

This chapter analyses this major shift in the history of environmental ideas from the vantage point of the Finnish Green party.[4] Traditionally, the history of environmental ideas has been looked at from an eco-philosophical perspective, analysing the thoughts of eco-philosophers, or focusing the analysis on a more grassroots level, such as the environmental movements themselves. While such approaches have value, the examination of how environmental ideas have been put to practice at the level of politics supplements this view while deepening our understanding of the practical implications of environmental ideas. After all, such scholars as Joachim Radkau and William Connolly have noted how ecological ideas have not become visible at the level of practical implementation despite their presence in public political discourse. It is thus worth scrutinising more closely the potential reasons for this lack of implementation.[5]

The Finnish Green party, a middle-sized party in the Finnish multi-party parliamentary system, provides a particularly interesting example of this development. The changes in the Finnish Greens' thinking can also be understood as a case study of a wider phenomenon: the transnational changes in practical implementation of environmental ideas. The Greens first embraced the social movements' radical environmentalist ideals in the late 1980s but, just a few years later, started leaning into market-friendlier environmentalism. By 1995, they had changed their ideological presuppositions thoroughly and, largely due to this, managed to become the first Green nation-level governmental party in Europe.

In its second section, this chapter demonstrates how the Finnish Green party drew ideas from radical environmentalism. Consequently, the Greens wanted to re-conceptualise many presuppositions regarding modernity. These included defining wellbeing in a materialistic manner through the growth of consumption, as well as a midset of mastery over nature that had guided human economic activity and led to the overconsumption of natural resources, as well as human estrangement from a more communal way of living. During the 1990s, the Greens redefined their environmental ideologies through market-friendly and growth-oriented concepts, such as sustainable development or eco-efficiency. Endorsing cooperation with the established political and economic

4. This shift is more deeply analysed as part of my forthcoming dissertation, see Matero 2023. Here, however, the vantage point is studied through the lens of environmental history as William Connolly's model of cultural belonging is applied as an analytical tool. This chapter thus not only demonstrates some of the key themes of the dissertation in question but also provides a particularly environmental historical analysis of it. See also Connolly 2017.

5. Radkau 2005, p. 294; Connolly 2017.

institutions became the new goal of environmental politics. Environmental political concepts were now derived from economic language, and alternative visions for society were replaced with ideals of green growth.

This chapter aims to understand the argumentation behind the turn to green growth ideals in order to see what the reasons behind these changes were for the actors. In the third section, I investigate this turn from the perspective of conceptual analysis, where the choice to use one concept over another becomes a political act.[6] It turns out that this ideological change was preceded by an intense contest over reconceptualising the meaning of greenness, as moderate reformist actors brought forth new, competing ideas of how environmentalism should be (re-)defined ideologically. Using newspapers, party pamphlets and party conference minutes as sources, I pay attention particularly to premises regarding wellbeing and its connection to growth-orientation, the often-demonised antagonist of environmentalism. This opens up views on complex ideological questions about how to relate to anthropocentrism and social traditions while endorsing environmental wellbeing, questions that the Greens tackled as part of their political ideological development into a normalised party functioning efficiently within established political structures. From this perspective, the turn to green growth appears as an ideologically significant rift from earlier environmentalism, and was partially caused by transnational development, such as the new political hegemony of economic competitiveness in post-Cold War Europe.

Furthermore, I demonstrate how this moderate reformist turn of the 1990s can be understood through the analytical lens of what social scientist William Connolly calls 'cultural belonging'. Connolly points out how environmental or ecological interventions in business-as-usual politics have typically become dragged down by a deep-rooted human drive for cultural belonging, which is also a prerequisite for meaningful action. Connolly brings attention to a fundamental need for a 'sense of layered fit between self and the world' that seems to guide the possibilities for the kind of action we tend to perceive as meaningful – action that does not drift too far away from the hegemonic premises and models of thought of the cultural paradigm. We might think radically, but in our actions we tend to adapt to our surroundings in order to render our actions meaningful.[7] Using Connolly's conception of 'cultural belonging' helps explain the paradox the Greens have faced, namely the implicit contradiction

6. Theory for analysing concepts as political acts – see Skinner 2002. How the meanings of concepts are changed in time – see Koselleck 2004; Bevir 1999.

7. Connolly 2017, 9-11, 81.

between radical interventions in the current paradigm, and the need to belong in the system from which one operates in order to be effective in one's actions. It also provides a potential explanatory framework for one of the big questions in the study of environmental history of ideas: why has the half-century-long discussion on environmentalism not led to improving environmental conditions in practice?[8] The moderate 'reformist' Greens explicitly expressed a need to act more efficiently within the system. I argue that this need for meaningful activity thus drove the change towards moderate growth-oriented thinking – towards a state of cultural normalcy, a state of market orientation, with the previously-detested premises of Western culture now accepted as a given. This change thus provides a more general example not only of how environmental discourses in general have developed over the past decades, but also of how easily radical ideas get drawn back into the contemporary paradigm of thought, particularly on the level of implementation.

Conceptualising Green environmentalism on programmatic level

Since the late 1970s, the Finnish environmental movement had united a wide range of alternative social movements under the unifying aim of formulating alternative visions for society and questioning the materialistic culture of growing consumption, particularly its destructive effect on the environment. The movement united nature protectionists, fair trade advocates, nuclear power protestors, vegetarians and eco-feminists, among others.[9] These different groups gathered in 1979 to protest about the planned draining of a small Koijärvi lake in Southern Finland, threatening rare bird species in the area. The activists were not only successful in their attempts to save the lake in question; the event also provided both a symbolic and a practical starting point for the Green political movement in Finland.[10] The Finnish Greens had their first MPs from 1983 onwards. The Green Alliance (Vihreä Liitto) was not registered as a party until 1988, though. Until then, the Greens had believed they would represent the grassroots movement better without a formal party structure, uniting grassroots movements under the umbrella of an alternative vision of society.[11]

---

8. Such a question is addressed e.g. by Radkau (2005, p. 294), for whom we have lived in an'Age of Ecology' for the past half-a-century on the level of environmental discussion while environmental conditions have kept deteriorating.
9. Mickelsson 2007, pp. 250–51; Paastela 1987, p. 15.
10. Paastela 1987, pp. 22–24, 56–60; Aalto 2018, pp. 106–09.
11. Välimäki and Brax 1991, p. 33; Sohlstén 2007, pp. 41–47, 53–55.

In their first two party programmes from 1988 and 1990, the Greens presented their demands to restructure not just the economy, but the entire Western value system supporting economic growth. This included the detested mindset of 'mastery' over nature and people, and the social and ecological unsustainability and inequality that followed from this mindset. Simply put, the Greens demanded a 'fundamental re-evaluation of the premises of Western culture'.[12] A materialistic conception of wellbeing was considered to be the cause of both social and environmental problems of society, as wellbeing was measured merely with increased consumption: the Greens described how the storytellers of old cottages had been removed to retirement houses and replaced with televisions, raising the question of whether increase in material living standards had actually increased wellbeing at all. Instead of material growth, wellbeing needed to be sought in spiritual and social needs. This re-evaluation of our culture was termed the Companionship Movement (kumppanuusliike), as opposed to the mindset of mastery over nature and over other humans too, as evinced in capitalist and socialist societies.[13]

In replacing mastery with companionship, the Greens avoided separating the needs of the environment and the needs of humans: when nature was no longer a mere resource, the wellbeing of humans and non-human life could be redefined as interconnected. As the demand of material growth was removed from the equation, wellbeing would also provide opportunities for more real forms of human wellbeing through, for example, deeper communal connections and self-realisation instead of consumption. The Green vision was to 'atone for the broken relationship between human and nature' by finding starting points for interaction between peoples, cultures and all life. The theme of companionship was thus to be understood through a deeper sense of holistic interaction and interconnections instead of competition and mastery.[14] In economic life, this meant that ecological carrying capacity should determine limits for individual ownership, as environmental needs needed to be prioritised over protection of property.[15]

The aforementioned radical standpoint based on companionship and growth-criticism had virtually disappeared from discussion by the mid-

---

12. Green Alliance Party Programme 1990, p. 1.
13. Green Alliance Party Programme 1990, Prologue, 1, 2; Green Alliance General Programme 1988.
14. Green Alliance General Programme 1988; Green Alliance Party Programme 1990, Prologue, 1; Vihreä lanka 16.6.1989, pp. 1, 5.
15. Green Alliance Party Programme 1990, pp. 2–5; Woods 2010.

1990s, however. In the 1994 party programme, the Greens noted how it was now time for a green market economy, and explicitly feared a decline in the gross-national product if other countries were to get there first, basing their argumentation on the utility of greenness from a competitiveness perspective. Finns could secure 'better positions in global economic competition in the future' if environmentalism were merged with cleaner technology exportation. The Greens' political goal had turned from revisioning premises of Western society to greening them. Conceptually, this market-friendly change towards greener growth was accomplished in Finland by adopting the new key concept of sustainable development (kestävä kehitys), which would be developed further in upcoming programmes.[16]

The concept of sustainable development had become widely known by 1987, through the Brundtland Commission, named after Norwegian social democratic prime minister Gro Harlem Brundtland.[17] Having a social-democratic background, the concept merged the goals of growth, social equality and environmental wellbeing, and emphasised global justice in 1987. The emphasis on green growth and green consumption was only later underlined by, for example, the European Commission's interpretation, demonstrating how the meanings of concepts are debated and can change over time.[18] The Green Alliance parliamentary group had criticised the concept of sustainable development in 1988: there would be even bigger ecological problems ahead if economic growth were increased, they claimed.[19]

By the mid-1990s, however, the Greens had changed their minds, as they embraced the market-friendlier version of environmentalism through sustainable development. While earlier emphasising, for example, natural protection issues, they now wanted to place emphasis on affecting 'millions of consuming choices' by ecological taxation, eco-labels and voluntary market-guidance for corporations in their new 1994 programme. This would generate green growth through increased consumer demand for eco-friendlier products. Consequently, environmental responsibility was re-allocated to individual consumers. Though it is beyond the time scope of this chapter, it is worth pointing out that, after

16. Green Alliance Party Programme 1994, p. 3.

17. Laine and Jokinen 2001, pp. 64–65; Dryzek and Scholsberg 2005, pp. 257–58; Dryzek 2005, pp. 145–48; WCED 1987.

18. See e.g. Knill and Liefferink 2007; Collier 1998; European Commission 1993/2005; European Commission 1995. The theme is further investigated by Matero & Arffman 2023 (forthcoming).

19. Vihreä Lanka 9/1988, pp. 1, 6.

spending seven years in government, the 2002 Green programme was even more focused on creating growth markets for environmentally friendly production. Eco-efficiency had become the new practical measure to be used for the greening of economic growth, as the idea of decreasing the amount of energy and natural resources used per produced unit started to seem possible with advancing cleaner technology.[20] Intriguingly, the notion of 'rebound effect' – that resource efficiency would increase overall consumption due to lowered production costs – was entirely lacking from public discussion at the time, even though the concept was well known in academic discussion by then.[21]

Most of these concepts stemmed not from the radical environmentalism of the 1970s as the earlier goals of companionship had done,[22] but rather from the environmental economics discourse of the same era. Environmentally aligned economics were developed to generate ideas to tackle environmental issues without having to question basic assumptions of modernity such as the growth paradigm. This provided the environmental debates with a much more moderate conceptual cluster of ideas. The developers of such ideas and concepts were typically economists rather than environmentalists.[23]

Over the 1990s, the Greens' understanding of wellbeing and human relationship with non-human nature thus changed dramatically. Having earlier derived their conceptualisations of these ideas from radical environmental grassroots movements, the Greens were now drawing concepts from moderate sustainability-discourse. Meanwhile, restructuring basic premises of Western culture was no longer in their political toolbox, as wellbeing was now understood within the framework of green growth and consumption, bringing the Green party discussion back to normalcy, within the cultural norms of economic vocabulary. It is worth looking at the Green public debates taking

20. Green Alliance Programme of Principles 2002, *Kestävä kehitys todeksi*; Green Alliance Party Programme 1990, p. 2. As for *eco-efficiency*, see also Schmidt-Bleek 2000, which was widely discussed among the Finnish Greens.

21. Akenji 2019, pp. 1–2; also addressed by Schmidt-Bleek (2000) who influenced the Finnish Greens.

22. The background of the *Companionship movement* can be traced back to ecofeminist philosopher Riane Eisler, whose book *Chalice and the Blade* was translated into Finnish by Green MP Satu Hassi. See Eisler 1988.

23. Warde et al. 2018, pp. 69–70; Caradonna 2019, pp. 155–57. The conceptual cluster of environmental economics did not automatically promote business-as-usual growth; e.g. Friedrich Schmidt-Bleek (2000) warned that it would only work in a more prudent value system. Nevertheless, these concepts were used in political discussion within the framework of green growth.

place simultaneously to understand how, why, and under what contexts these changes took place. It turns out that the need for 'cultural belonging' – a need for efficient action in collaboration with the rest of society – was at the core of this ideological change.

Arguments behind the change

The changes above need to be understood within a wider domestic and international context. Changes were taking place in the early 1990s, with the end of the Cold War, and a harsh depression, with mass unemployment hitting Finland, causing the growth-critical ideals of the Greens suddenly to seem questionable. In Central Europe, media emphasis was on the unification of the Germanys. A power shift was taking place within the most famous Green party in Europe, the German Green Party, who declared themselves a 'reformist' party in 1991. This change, led by political leader Joschka Fischer and reformist thinker Hubert Kleinert, was partly due to a devastating election loss in 1990. Blame was placed on a radical programme and uncompromising refusal to cooperate with the established political and economic order. Being a trustworthy party in the eyes of the electorate had become a new premise for reforming the radical environmental German Greens into a reformist civil rights party that could even participate in government work.[24] Hubert Kleinert pointed out a contradiction between 'efficiency' and 'legitimacy' in green radicalism: as efficient ways to make politics through parliamentary participation required ideological compromises and cooperation, radicals did not consider such means to be legitimate.[25]

This was the context the Finnish Greens faced in the early 1990s. They gained massively increased support in the 1991 elections with their radical programme, raising their seat number in the 200-seat Finnish parliament from four to ten. Unlike their German counterparts, the Finnish Greens had in fact gained more votes with their radical growth-critical thinking, despite the depression. However, they ran into trouble when participating in government negotiations: it was soon made clear by bigger parties that no growth-critical party could make it into government during a depression. Reformist ideas had been well known among Finnish Greens but, after 1991, they were for the first time supported by party leadership, partially because reformist Pekka Sauri became party Chair and started guiding the party towards a more moderate

24. Die Grünen 2019, pp. 44–47; *Der Spiegel* 50/1990, pp. 28–29.

25. Die Grünen 2019, pp. 44–47; *Der Spiegel* 23/1991, p. 35.

direction that would perhaps enable participation in governmental work. Chief ideologist Osmo Soininvaara demanded politics without dreams (haavetonta politiikkaa), because 'all-or-nothing politics would result in the outcome of nothing', following in the efficiency-oriented footsteps of the German Greens. With Finland plunging into depression and the Greens remaining involuntarily in opposition, anti-growth idealism understandably started to seem outdated for many moderately-minded Greens.[26]

A couple of years before these 1991 events, discussion on reformism had already started to take shape in the pages of *Vihreä Lanka*, the independently-run newspaper of the Green Party. In 1988, the most widely discussed issue in the newspaper was traditional forest and water protection questions. In 1989, reformist ideals of market-friendly green consumerism suddenly got wider attention. New Chief Editor Pauli Välimäki hoped to keep ideologically pluralistic discussion present in the pages of the magazine, consequently giving more space to the reformist opposition, while himself often calling for more pragmatic approaches to Green politics in his editorials. German Reformist leader Joschka Fischer was quoted demanding the Greens to appeal to more middle-class voters.[27] Osmo Soininvaara was especially often quoted in the magazine among Finnish reformist thinkers, demanding for example environmentalism that utilised 'the power of market forces'.[28] He also supported the idea of correcting market flaws of overconsuming resources with proper price control.[29]

These ideas raised immediate controversy: for Green thinker Olli Tammilehto, green language was being overrun by market ideology. Social needs and a real search for meaning were being replaced with hollow materialism, he thought, as the Greens were entering the paradigm of 'our system'. He also saw new tension rising between the party and the environmental movement, that the party supposedly represented. The so-called reformists of the party were aiming at more effective action through parliament, which practically meant 'shutting out the citizens' movement from the map'.[30] There were understandable reasons for Tammilehto's concerns: Osmo Soininvaara had noted at the 1989 Green party spring conference how the party should move towards a reformist rather than radical direction. *Vihreä Lanka* magazine went so far as to claim that the Greens were indeed developing into a 'reform party' – although no

26. *Vihreä Lanka* 21 Mar. 1991, p. 2.
27. *Vihreä Lanka* 5/89.
28. *Vihreä Lanka* 8/89, p. 2; *Vihreä Lanka* 12/89, p. 2.
29. *Vihreä Lanka* 7/89.
30. *Vihreä Lanka* 27/89, p. 2; *Vihreä Lanka* 2/89, p. 4.

such statement was made on a programmatic level. Green Party chair Heidi Hautala fiercely opposed this claim, calling the Green movement an inherently anti-capitalist one, while ridiculing the naïve progress-believers for thinking that development could ever be truly sustainable within this kind of system. Hautala was not willing to join the celebration around the German reformist Joschka Fischer either. For her, society needed 'no growth of gross-national product, but growth of political imagination'.[31] As this discussion demonstrates, the ongoing German Green discussion was also closely followed in Finnish discussion, giving transnational dimensions to the reformist debate.

The new moderate green ideology was represented in the Green ABC Book 2 (Vihreä ABC-kirja 2), published in 1995. The first Green ABC Book had demonstrated radical green ideals a few years earlier but, in the second, the Greens were seeking compromises with the prevailing culture, beliefs and institutions. The Greens set out to become a party among others that represented a larger electorate than merely the alternative movements. They wanted to appeal particularly to young well-educated voters of the cities who were known to readily vote Green.[32]

The head of the argumentative spear was thus aimed against the ineffective outcome of radicalism: the Greens needed to get legislation done, cooperate with institutions in order to make it to government, and create credibility for voters. The similarity in argumentation with the German Green reformists in striking. These themes were encapsulated in the argumentative tool of 'maturity', presented time and again in Green ABC Book II. By setting pragmatic political goals (instead of ideological ones) and thus overcoming the 'children's disease' of radicalism, the Greens had now 'matured' for many of the reformists who were thus rewriting the story of Green party development from naïveté to responsible adulthood. The antithesis of this newly-found maturity was the 'naïve' innocence of youthful radicalism. The critique of radicalism was thus not thoroughly critical, as such innocence had its place when growing up, but it nevertheless did not belong in the lives of adults; there was a hierarchy and perhaps a sense of condescension present in the reformist perception of radicalism.[33]

31. Green Alliance Spring Conference Minutes 1989, §5; *Vihreä Lanka* 22/89, p. 2; *Vihreä Lanka* 37/89, p. 7.

32. Välimäki 1995, p. 73, in *Green ABC Book 2*; see also Aalto 2018, pp. 377–81; Remes 2007, pp. 91–92; Isotalo 2007, p. 143; Ylikahri 2007, pp. 186–88.

33. Välimäki 1995, pp. 69–71, 117–18.

At the 1994 party conference, the new programme draft had by then been presented by former chair of the party, Pekka Sauri. Conference minutes reveal that Sauri presented the programme as one that was 'structured to bring the entities of economy and ecology closer together'. A more sustainable direction for society would thus be achievable with the methods of the market economy.[34] A debate followed, but no major changes were made to Sauri's draft proposal. The ideological change to reformism was thus finalised on a programmatic level after roughly five years of debating. A year later, the Finnish Greens joined the national government as the first Green party in Europe, providing evidence of the utility of reformism.[35]

The ideals of green growth were thus deployed as a political tool to get things done, and to change the Green course away from the ideological foundations of the radical grassroots environmental movements. Practical sacrifices were necessary in order to become efficient in the political game. In the words of Pekka Haavisto, who became the first Green minister in Finland, they would not have got anything done with the attitude of the environmental movements.[36] For the radical anti-modernists, meanwhile, parliamentary politics was only another forum where the ideals and new foundations of an alternative future could be laid, with little attention paid to actual legislative processes. Giving up on those ideals was considered morally questionable, even wrong. Former party vice chair Ulla Klötzer called it a betrayal. Some – like Klötzer – left the party altogether, but others stayed and continued to work from within the now-reformist party. Martti Lundén, for example, became one of the founders of the Eco-Greens Association in 1995 to serve as an ecological opposition from within the party after becoming disappointed with the increased anthropocentric emphasis. This dissident association still exists.[37]

There were thus two stories in play here. The reformist story was one of maturity, where responding to pragmatic needs with green growth ideals would lead to more efficiency; the radicals meanwhile saw these changes as selling out the ideals that the Greens were meant to represent, that is, the radicalism of the alternative movements. A materialistic conception of wellbeing based on economic growth was either critiqued and re-defined in a more holistic way

34. Green Alliance Party Conference Minutes 1994. 11–12 June 1994, §8–9.

35. The German Greens had already been a state-level government party in the 1980s, though.

36. Isotalo 2007, pp. 162, 179.

37. Helsingin Sanomat 28 May 1998; Lunden 2019.

(as the radicals did) or accepted as a political reality that needed to be adapted to in order to cooperate efficiently within the system (as the reformists did).

Retreating from green radicalism to within the established growth paradigm can be understood as a case study of a larger phenomenon that has been typical for environmental discourses. Connolly's notion of 'cultural belonging' as prerequisite for meaningful action seems particularly believable as the ineffective radical standpoints faded from discussion altogether. They would, of course, have been impractical obstacles to effective action within the paradigm in which the Greens felt it necessary to operate. As mentioned earlier, attempts at radical paradigm shifts have typically been pulled back by the need for cultural belonging – that is, locating oneself in such a way in relation to contemporary institutions that meaningful action becomes possible, even at the possible expense of urgent paradigmatic changes. Thus, the Finnish Greens become an example of a wider, global phenomenon – one which, for example, the aforementioned German Greens had already faced before their Finnish counterparts.

Along with the influence of Die Grüne, the end of the Cold War and the collapse of the Eastern Bloc provide context for these changes. The Zeitgeist had changed in just a couple of years, claimed the German newspaper *Der Spiegel*, and capitalism had emerged victorious.[38] However, even without the collapse of the Eastern Bloc, consumerism and simplified regulative practices were a rising global political theme due to increased globalisation and increasing emphasis on economic competitiveness. Even in the EU, strict environmental standards had already caused bigger costs for production, which meant disadvantages in economic competition. As Knill and Liefferink point out, the themes of sustainable development were used as a conceptual tool at the EU level to replace harsh industry regulation that was hurting industry competitiveness in a globalising world. Member states soon found themselves in a regulatory competition to create favourable competitive conditions for industries, and this meant reallocating environmental responsibility to the consumer.[39]

There was thus harsh pressure for the Greens to adapt to the competitiveness paradigm by the early 1990s – to a 'hegemony' of neoliberalism, as conceptual historian Niklas Olsen has called the so-called 'market democracy', where regulation is loosened and responsibility reallocated to consumers in order to enhance competitiveness. Paradoxically, state regulations that had earlier protected the consumer against market forces were now perceived as limits

38. *Der Spiegel* 50/1990, pp. 28–29.

39. Knill and Liefferink 2007, pp. 103, 163–64; See also Collier 1998, p. 25.

to consumer agency.[40] Greenness needed to be adapted to this new political paradigm through the ideals of green growth if the Greens were to participate in government under this framework. It needs to be noted, though, that the reformist Greens would certainly not have considered themselves part of this hegemony of neoliberalism, as Olsen calls it, but nevertheless felt compelled to embrace the new standard for practical reasons, adapting their ideology to the prevailing political realities. Thus, when the growth-critical discussion of companionship faded from Green public discussion, the Finnish Greens were indeed taking part in a transnational turn, adapting to the new paradigm of global competitiveness.

Conclusions

Originally, the Finnish Green party ideology had originated from the environmentalist discourses of alternative movements. As changing political realities – such as the hegemony of consumeristic politics – compelled them to adapt, reformist-led Greens of the 1990s gave up on these supposedly naïve ideals. Nature was again something that needed to be used for resources to ensure human wellbeing, guided by such political practices as eco-labelling that would help consumers choose more environmentally-friendly products in the free market, while not hurting industry competitiveness in the process. From the perspective of the history of ideas, this can be understood as a major turn in Green party environmental thinking, as the mastery mindset was no longer critiqued nor was a deeper sense of companionship with nature demanded.

This change can be understood in terms of pulling back to normalcy, caused by the need for cultural 'belonging' and efficient action in the terms of William Connolly. The radicals had been faced with a crippling inability to 'get things done' – that is, to break free from the constraints of anthropocentrism and socio-centrism on the level of institutional action while maintaining efficiency in the political field. Eventually, the Green party in Finland solved this paradox between efficiency and legitimacy by drawing concepts from an entirely different tradition of thought, that of environmental economics, helping them turn their environmentalism towards a more moderate, culturally acceptable direction. Thus, the reformists were bringing the Greens to the ideological sphere of 'cultural belonging' and consequently marginalising the more radical environmental language. It remains an intriguing question (although an unanswerable one) whether both forms of greenness could have co-existed in

40. Olsen 2019, pp. 251–55, 261–63.

political discourse, had the radicals perhaps been less eager to fundamentally reject cooperation, or had the reformists been more willing to maintain long-term radical environmentalist ideals, despite making compromises on the short term.

This division of thought between radicalism and reformism has been articulated either as 'naïve adolescence' that will not accomplish anything practically, or as 'selling-out' of key green principles on the all-consuming altar of the growth economy, thus giving up the alternative vision of the future that the Greens had longed to represent. While understanding both sides of the debate, I believe it is worthwhile to look beyond these simplistic forms of argumentation. For an environmental historian, the interesting question should be focused on overcoming the obstacles that social movements seem to face when dealing issues that need to be addressed quickly and often radically: how can an environmentally legitimate paradigm shift be advanced without either operating from within the destructive practices, or resorting to ineffective radicalism that has little to no effect? How is this paradox to be tackled without such immense sacrifices one way or another?

I opened this chapter by asking why ideas from the environmental movements seem to have trouble getting implemented. The case example of the Finnish Green Party demonstrates how difficult it is to bring new controversial ideas to established structures such as party politics without having to make compromises in order to find meaningful ways to 'belong'. From a political perspective, it made a lot of sense: after all, the doors to government were opened to the Greens by such decisions. When looking at these ideological changes through the lens of environmental history, however, they become yet another example of why and how environmental ideas tend to get dragged down to anthropocentrist premises, at worst even devoured by economic interests. Party politics thus becomes a prime example of this drive towards cultural belonging that we, as humans, all possess.

Bibliography

Aalto, Sari. 2018. *Vaihtoehtopuolue: Vihreän liikkeen tie puolueeksi.* Helsinki: Into Kustannus.

Akenji, Lewis. 2019. Avoiding Consumer Scapegoatism: Towards a Political Economy of Sustainable Living. Doctoral Dissertation. University of Helsinki.

Bevir, Mark. 1999. *The Logic of the History of Ideas.* Cambridge: Cambridge University Press.

Caradonna, Jeremy L. 2018. 'An incompatible couple: A critical history of economic growth and sustainable development'. In Iris Borowy and Matthias Schmelzer (eds), *History of the Future of Economic Growth. Historical Roots of Current Debates on Sustainable Degrowth.* London: Routledge.

Collier, Ute. 1998. 'The environmental dimension of deregulation: An introduction'. In Ute Collier (ed.) *Deregulation in the European Union: Environmental Perspectives*, pp. 3–22. London: Routledge.

Connolly, William. 2017. *Facing the Planetary: Entangled Humanism and the Politics of Swarming.* Durham, NC and London: Duke University Press.

Daly, Herman. [1977] 1991. *Steady State Economics.* 2nd Edition. Washington D.C.: Island Press.

Der Spiegel newspaper issues: 50/1990; 23/1991.

Die Grünen 2019. Zeiten Ändern Sich. Wir Ändern Sie Mit. Bündnis 90 / Die Grünen. https://cms.gruene.de/uploads/documents/GRUENE_Chronik_1979-2019.pdf

Dryzek, John and David Schlosberg (eds). 2005. *Debating the Earth: The Environmental Politics Reader.* Oxford: Oxford University Press.

Dryzek, John. [2005] 2013. *The Politics of the Earth: Environmental Discourses.* Oxford: Oxford University Press.

Eisler, Riane. 1988. *Malja ja miekka: Historiamme ja tulevaisuutemme.* Helsinki: WSOY.

European Commission 1993/2005. 'Towards Sustainability': the European Community Programme of policy and action in relation to the environment and sustainable development, *Official Journal of the European Communities C138.*

European Commission 1995. Comments of the Commission on the Report of the Independent Experts Group on Legislative and Administrative Simplification. SEC95 2121 Final.

Freeden, Michael. 2006. 'Ideology and political theory'. *Journal of Political Ideologies* **11** (1): 3–22.

Fressoz, Jean-Baptiste and Christophe Bonneuil. 2018. 'Growth unlimited. The idea of infinite growth from fossil capitalism to green capitalism. In Iris Borowy and Matthias Schmelzer (eds), *History of the Future of Economic Growth. Historical Roots of Current Debates on Sustainable Degrowth.* London: Routledge.

Green Alliance Programme of Principles. 2002 (*Vihreän Liiton periaateohjelma*). POHTIVA Archive: https://www.fsd.tuni.fi/pohtiva/ohjelmalistat/VIHR/855

Green Alliance Party Programme. 1990 (*Vihreän Liiton puolueohjelma*). POHTIVA Archive: https://www.fsd.tuni.fi/pohtiva/ohjelmalistat/VIHR/884

Green Alliance Party Programme. 1994 (*Vihreän Liiton puolueohjelma*). POHTIVA Archive: https://www.fsd.tuni.fi/pohtiva/ohjelmalistat/VIHR/885

Green Alliance General Programme. 1988 (*Vihreän Liiton yleisohjelma*). POHTIVA Archive: https://www.fsd.tuni.fi/pohtiva/ohjelmalistat/VIHR/883

Green Alliance Party Conference Minutes. 11–12 June 1994.

Green Alliance Spring Conference Minutes. 27–28 May 1989.

Helsingin Sanomat newspaper issues: 28 May 1998.

Hockenos, Paul. 2008. *Joschka Fischer and the Making of the Berlin Republic: An Alternative History of Postwar Germany*. Oxford: Oxford University Press.

Isotalo, Merja. 2007. 'Hallituspuolueeksi EU-maahan & Vihreät vallan kahvassa'. In Tanja Remes, Tanja and Jemina Sohlstén (eds), *Edellä! Vihreä Liitto 20 vuotta*. Helsinki: Vihreä sivistysliitto.

Knill, Christoph and Duncan Liefferink. 2007. *Environmental Politics in the European Union: Policy-making, Implementation and Patterns of Multi-level Governance*. Manchester: Manchester University Press.

Koselleck, Reinhart. 2004. *Futures Past: On the Semantics of Historical Time*. New York: Columbia University Press.

Laine, Markus and Pekka Jokinen. 2001. 'Politiikan ulottuvuudet'. In Yrjö Haila and Pekka Jokinen (eds), *Ympäristöpolitiikka*. Tampere: Vastapaino.

Lundén, Martti. 2019. Alkuvaiheet: https://www.ekovihreat.fi/yhdistys/alkuvaiheet/ Accessed 22 Sept. 2019.

Matero, Risto-Matti and Atte Arffman. 2024. 'An economic tail wagging an ecological dog? Well-being and sustainable development from the perspective of entangled history'. In Kortemäki et al. (eds), *Interdisciplinary Perspectives on Planetary Well-being*. Oxon & New York: Routledge. Forthcoming.

Matero, Risto-Matti. 2023. From Companion with Nature to Green Growth. Competing Conceptualisations on Well-being and the Environment in Finnish and German Green Parties, 1980–2002. Doctoral Dissertation. University of Jyväskylä. Forthcoming.

Mickelsson, Rauli. 2007. *Suomen puolueet: historia, muutos ja nykypäivä*. Tampere: Vastapaino.

Milder, Stephen. 2019. *Greening Democracy: The Anti-Nuclear Movement and Political Environmentalism in West Germany and Beyond, 1968–1983*. Cambridge: Cambridge University Press.

Naess, Arne. 1997. 'Pinnallinen ja syvällinen, pitkän aikavälin ekologialiike'. In Markku Oksanen and Marjo Rauhala-Hayes (eds), *Ympäristöfilosofia*. Helsinki: Gaudeamus.

Olsen, Niklas. 2017. *The Sovereign Consumer: A New Intellectual History of Neoliberalism*. Copenhagen: Palgrave Macmillan.

Paastela, Jukka. 1987. *Finland's New Social Movements*. Tampere: Tampereen yliopisto.

Radkau, Joachim. 2005. *Nature and Power: A Global History of the Environment*. Cambridge: Cambridge University Press.

Radkau, Joachim. 2011. *Die Ära der Ökologie: Eine Weltgeschichte*. Münich: Verlag C.H. Beck.

Remes, Tanja. 2007. 'Kohti "tavallista puoluetta"?' In Tanja Remes and Jemina Sohlstén (eds), *Edellä! Vihreä Liitto 20 vuotta*. Helsinki: Vihreä sivistysliitto.

Schmidt-Bleek, Friedrich. 2000. *Luonnon uusi laskuoppi: ekotehokkuuden mittari MIPS*. Helsinki: Gaudeamus.

Skinner, Quentin. 2002. *Visions of Politics I: Regarding Method.* Cambridge: Cambridge University Press.

Sohlstén, Jemina. 2007. 'Liikkeen voima & Vihreää valoa puolueelle'. In Tanja Remes and Jemina Sohlstén (eds), *Edellä! Vihreä Liitto 20 vuotta.* Helsinki: Vihreä sivistysliitto.

Taylor, Bron. 2010. *Dark Green Religion: Nature Spirituality and the Planetary Future.* Berkeley: University of California Press.

Warde, Paul, Libby Robin and Sverker Sörlin. 2018. *The Environment: History of the Idea.* Baltimore: Johns Hopkins.

WCED. 1987. Our Common Future: Report of the World Commission on Environment and Development. United Nations.

Vihreä Lanka newspaper issues: 9/1988, 2/1989, 5/1989, 7/1989, 8/1989, 12/1989, 22/1989, 27/1989, 37/1989, 21 March 1991.

Woods, Kerri. 2010. *Human Rights and Environmental Sustainability.* Cheltenham: Edward Elgar.

Välimäki, Pauli and Anne Brax (eds). 1991. *Vihreä ABC-Kirja.* Helsinki: Vihreä Liitto.

Välimäki, Pauli (ed.) 1995. *Vihreä ABC-Kirja 2.* Helsinki: Vihreä Liitto.

Chapter 4

THE CHANGING STATUS OF BIRCH TREES IN FINNISH FORESTS FROM THE SEVENTEENTH CENTURY TO THE TWENTIETH CENTURY

Seija A. Niemi

Introduction

Forests are 'Green Gold', the highly valued treasures of Finland. For centuries, the livelihood of the Finns has been based largely on forest products, both in domestic households and in foreign trade. In the ancient times, among the main export articles were furs and pelts, then tar and timber and, in the past two centuries, paper, pulp and timber.

In Finland, forestry is a well-examined topic, as are the ecology, biology and geography of Finnish forests.[1] However, the history of forests has only recently received wider attention. Among the first environmental history studies on Finnish forests was my 2005 licentiate thesis, *Suomalaisen metsäluonnon lukemisen historiaa – Ihmisen ja koivun muuttuva suhde Suomessa 1730-luvulta 1930-luvulle* (The History of Environmental Literacy in the Finnish Forests – the Changing Relationship between Humans and the Birch Tree from the 1730s to the 1930s in Finland). In my study, I analysed some factors of environmental literacy, such as the power of dictating how people read their environment, and to what extent this power modifies or shapes environment. I used birch trees as my example of various ways of reading the language of forests. I also wrote a book, *Koivu – Suomen kansallispuu* (Birch – National Tree of Finland, 2016) based on my thesis.[2]

1. Vallinkoski 1963, pp. 146, 254; Hollsten 1996, p. 2; Emanuelsson 1997, p. 42; Michelsen 1999, p. 195; Ruuttula-Vasari 2004, p. 20.

2. A great deal of the information in this article is based on my thesis, Niemi 2005, and my book, Niemi 2015.

doi: 10.3197/63824846758018.ch04

After the Ice Age, the birch came as the first tree species to the area now called Finland.[3] If the forests had grown in their natural succession after the Ice Age, the longer living pine and spruce would have gradually exterminated birches. However, human interference, for example, when the ancient hunters burned trees to make clearings for hunting moose and other game, and later slash-and-burn cultivation, helped birch trees to survive. In recent centuries, the value of pine and spruce grew significantly as paper mills began to use these wood fibres in their production, while birch became an undervalued species since it had no such use. Gradually birch conquered a part of paper processing industry and became an appreciated tree there too.[4] Thus, birch has had its ups and downs, which makes it an interesting tree to study.

This article will illuminate the changing status of the birch tree, the ways in which Finns have perceived, used and valued it. It will also give some examples of the standards of Finnish environmental literacy from the seventeenth century to the twentieth century. This period covers both the pre-industrial and the industrial socio-economic changes in the environmental history of the birch tree.

Birch trees in the Finnish forests

The Northern Hemisphere is the realm of birches. Over 120 birch species, generally small to medium-sized trees or shrubs, grow in northern temperate and boreal climate zones. Birch trees are the biggest trees in Finnish forests. Their roots hold strong in ground even during storms. For successful growth, they need lots of light. It is difficult to breed birches from seeds: only 25–35 per cent of seeds will sprout.[5]

In Europe, Finland's forests are relatively the largest: of Finland's thirty million hectares, 26 million hectares, 86 per cent of the country's surface, are covered with the woods. The three most common tree species are pine, spruce and birch. The three main Finnish birch species are the downy birch (*Betula pubescens*), the silver birch (*Betula pendula*) and the dwarf birch (*Betula nana*). The most valuable are cultivars of silver birch. Another valuable form is the blaze birch.[6]

3. Myllyntaus 1999, p. 88.
4. Holopainen 1957, p. 32.
5. Tomppo 1998, 9.
6. Niemi 2006, p. 17.

Birch in Finnish is '*koivu*'. The original formative Uralic word is *kojwa*. In other Finno-Uralic languages birch is called for example '*koiv*' in Veps, '*kõiv*' or '*kask*' in Estonian, '*kue*' in Mari, and '*kő*' in Nenents.[7] In Sámi, birch is '*soahki*', in Swedish '*björk*', in Danish '*birk*', in German '*Birke*', in Russian '*berjoza*', in France '*boaleau*' and in Latin '*betula*'.[8]

In the old days, peasants called different kinds of birches by various names, mostly according to their growing place or their use. For example, in forests grew '*laulukoivu*' (song birch), '*rauvus-koivu*' (iron birch), '*lasi-koivu*' (glass birch), '*kääpiökoivu*' (dwarf birch), '*tunturikovu*' (fjell birch) and '*perjantaikoivu*' (Friday birch).[9] Birch also such names as '*the Tree of Wisdom*', since flexible birch branches were suitable for physical punishment. As late as the 1950s, a cane had the name '*the Master from the Birch Peninsula*' (Koivuniemen herra). Physical punishment was not yet forbidden in Finland at that time.

Environmental literacy

Charles E. Roth, an American scholar, defined *environmental literacy* in the 1960s: '[T]he capacity to perceive and interpret the relative health of environmental systems and take appropriate action to maintain, restore, or improve the health of those systems.'[10] In other words, a person is capable of perceiving harmful developments in his/her environment – that is, to 'read' the environment. A person is also able to react to destructive changes, either by preventing or remedying them - that is, to 'write' his/her own environment.[11]

In different periods people have perceived their environment in different ways based on their ways of thinking, conceptualising and valuing their surroundings. For example, in the seventeenth century, when Finland was a part of the Swedish kingdom, the greatest fear among Swedish mine owners was that common people would ravage the forests to extinction. They were not at all worried that they themselves ruined the forests around the mines by overcutting the trees.[12]

7. Häkkinen 2004, p. 456.
8. Kalm 1759, p. 3.
9. Ibid., pp. 4–5.
10. Roth 1992, p. 3.
11. Disinger and Roth 1992, pp. 4–5; Hares et al. 2006, pp. 5–6; Hsu and Roth 1998, pp. 229–49.
12. Niemi 2006, p. 17.

The common people lived in a close connection with their environment. Their awareness of the environment was based on their language, attitudes and values inherited from their ancestors. Their observations and experiences relied on inherited attitudes and values. Upon these inherited attitudes and values, they built their interpretation, definition and holistic understanding of nature.[13]

Different groups have different kinds of environmental literacy. Still today, cultural and socio-economic transitions affect our ability to interpret and understand the environment, as fast as the ecological changes take place. The values, ideologies and interests of decision makers influence the environment, as they use their power to make environmental decisions.[14]

Finns' traditional way of interpreting the environment

The first inhabitants on the area nowadays called Finland arrived when the ice cover gradually melted after the Ice Age. These newcomers were hunters following fur animals such as deer, moose, beaver, hare, fox and seal. Their environmental literacy emphasised knowledge of the surrounding world and understanding the regularities of different seasons. Hunters observed the life and routes of different animals, as hunters and gamekeepers still do nowadays. In the course of a year, hunters moved from place to place depending on where the best game, fish, plants and other natural resources were to be found. Presumably they wanted to achieve the same hunting results every year, but eventually, when the number of the people grew but the catch remained the same, unavoidable extinction of some animals occurred.

The impact of the ancient Finns on physical surroundings was almost invisible. However, some changes happened when they burned forests to make so called 'game meadows': open areas with tasty young forest growing after fire. Moose and deer eat young birches and other deciduous trees year-round. When these animals came to eat on the meadows, they were an easy target for hunters. Figures of moose and deer are common elements in ancient rock paintings, especially in south-eastern Finland. These figures can also be totem animals of various clans or groups.[15]

13. Orr 1992, p. 88; Golley 1998, pp. 1, 67–8.
14. Niemi 2018, p. 27.
15. Ruotsala 2002, p. 326; Niemi 2005, p. 12.

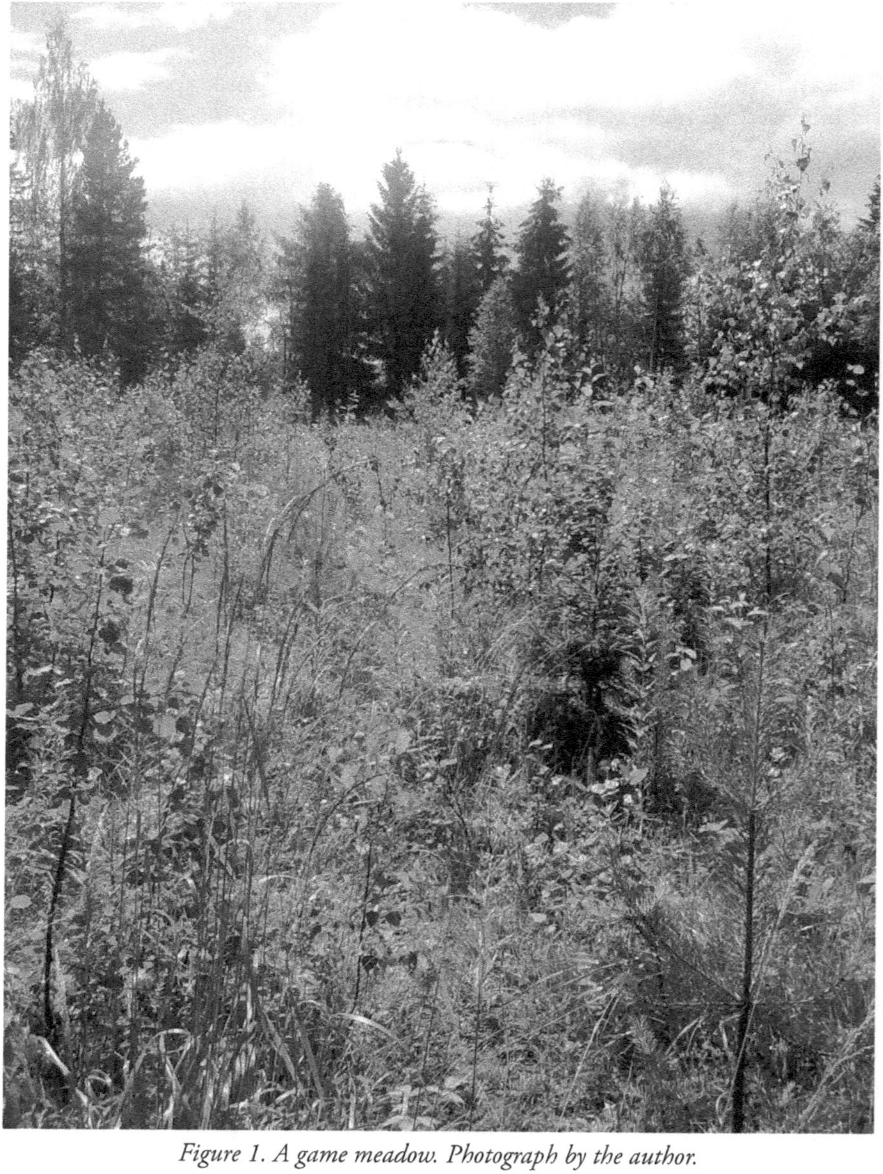

Figure 1. A game meadow. Photograph by the author.

The ancient hunters used birch for several purposes. First, it was good firewood; and it was also good material for parts of their weapons. Bows were essential equipment in a hunt. The ancient Finns made their bows from one or two tree species. Carl von Linné, who researched the Sámi people in the eighteenth century, found out that the Sámi people made the bows of pine and birch and fastened the different wooden pieces together with perch glue. Another wood combination was willow and birch, so that outer part of the bow was birch and the inside willow. The two components were stuck together with pitch and the bow was covered by a narrow birch bark band wound around it.[16]

In wintertime, snow covers Finland for months. In the ancient times, skis were the most suitable means of moving from place to place, since there were no roads or other suitable forms of transport. When the old Finns went hunting, they used their bow as a pole in one hand. In the other hand they might have had a lance. Finns used two different types of skis. Both skis could either be equally long or the right ski was shorter than the left. The equally-sized skis were usually made of pine as was the longer ski in the odd pairs. The shorter ski was made of deciduous wood, usually birch. Such skis often had fur underneath, especially the shorter one, so that the hair helped in moving forward and prevented slipping backwards.[17]

Excellent material for household equipment[18]

Eventually the hunting way of life of the old Finns changed to farming. Forests grew and still grow all over Finland; thus they are familiar places for the Finns. In the ancient times, forests surrounded homes and villages. Wherever people wanted to go, they had to go there through a forest. In the forests, people kept their eyes open and perceived suitable material for all kinds of needs. Forests were old time supermarkets where one could pick up the right material for every purpose imaginable.[19]

Peasants knew that birches grow on all kinds of soils. On poor, nutrient-poor and low-lying ground, birches may grow crooked, and the wood turns hard, tight and difficult to work with. On dry and nutritious soils birches grow straight and the wood will be easy to mould for different purposes. On the other hand, birches grow even on poor soils where no other trees manage to grow. Birches feel themselves at home with other tree species, but among spruces, pines and junipers they will gradually disappear.[20]

16. Vuorela 1998, pp. 26–31.
17. Ibid., pp. 678–80.
18. The information here about ancient ways of using birch is mostly based on Kalm 1759.
19. Niemi 2005, p. 72.
20. Kalm 1759, p. 6.

Figure 2. Solid birch wood. Photograph by the author.

Birch wood is white, smooth and beautiful. It is strong, hard and durable. In the old days, people made axles of wheels, parts for carriages, sleighs, skids, knife and spade handles, handles, sticks, and all kinds of agricultural equipment from birch. They turned from birch spindles, bobbins, various kinds of dishes, boxes, rattles, dolls and other toys. Even the bandits in Germany used robust birch sticks, with pieces of root left on the head, as travel cudgels, and common travellers used same kind of cudgels.[21]

In such regions where few or no spruce or pine trees grew, people used birch as building material as well. Birch keeps its beauty and durability in a dry place under a roof. But if it is left out for longer periods, it will rot quickly. In Russia, in some areas around Moscow, people even built bridges out of birch, but had to repair them constantly.[22]

21. Ibid., pp. 11–12.

22. Ibid., p. 6.

Every inch of birch is useful, even the ash

Every part of a birch is useful, from roots to top, big and small branches with or without leaves, inner and outer bark, buds, pollen, sap, and even the ash. One of the most suitable uses is warming up buildings with birch firewood.

Green birch leaves were tasty feed for cattle. Leaves can be collected in all kinds of places: in forests and on mosses, on islands and islets, in glades and meadows. Peasants cut off leafy branches or ripped the leaves off a tree if they did not want to chop it down. Branches with leaves were stored either one singly or in bunches called '*kerppo*' or '*kerppu*'. Cattle ate the dried branches and leaves as they were, or softened in hot water before use. The taste of birch leaves is best when they are collected shortly after midsummer. They taste bitter in the spring, and, in the autumn, turn to chewy like skin. Usually, men felled trees and women and children cut branches with billhooks. In archaeological excavations, billhooks are found from earlier periods than sickles and scythes. This indicates that the collecting of leaves is a very old habit and that people used to build their dwellings near deciduous forests.[23]

Birch leaves are still suitable for many purposes today. You can dye thread or textiles yellow if you put birch flowers and buds with lots of pollen in the liquid. Dried birch leaves are suitable for tasty herb drink or tea. The leaves contain plenty of diuretic ingredients as well as lots of proteins and vitamins C, B_2 and B_3.

Birch twigs and branches are suitable for many other purposes as well. In summertime, Finnish houses are sometimes decorated by fresh birch branches, especially when it is time to celebrate something, for example summer weddings, birthdays and anniversaries. At midsummer, it is a custom to decorate both sides of doors with birch branches. A sauna without a fresh birch '*vihta*' or '*vasta*' is not a proper sauna, according to many Finns. Silver birches provide the best material for winter vihtas. Downy birch is suitable for summer vihtas. A new vihta-year begins after midsummer. Then you burn the last '*talvi-vihta*' (winter-vihta) in your midsummer fire and make a first new vihta when you walk to your sauna which, of course, is beside one of the hundreds of thousands of lakes or rivers or ponds in Finland.

In the old days, people used the outer white birch bark for various purposes. They created beautiful and practical artefacts such as caskets, boxes, cases, containers, shoes, horns, rings and knapsacks. They made ropes, and covered stones to act as sinkers on fishing nets. Large waterproof and bendy

23. Ibid., pp. 21–22; Laurén 1987, p. 75; Slotte 1996, pp. 201, 203; Vuorela 1998, pp. 216–17; Niemi 2005, pp. 55–58.

bark plates were used to insulate roofs. These plates were covered either by logs or turf. If coated in tar, they lasted even longer. You can see these birch bark roofs on some old houses even today. Before paper was invented, people used to write messages on birch bark, especially in Russia. In England people used birch bark as lamps. Still today, the outer white birch bark is the best tool to ignite any kind of fire.[24]

Figure 3. An old Russian birch bark roof. Photograph by the author.

The brown bark between the outer white bark and the wood inside a birch was used to tan leather. With this bark people also tanned fish nets since fish swim more readily into dark nets. Brown bark is good fertiliser for flower beds. In hard times when the crops failed people dried this bark, ground it and added it to flour. The Sámi people as well as Native American tribes used dark birch bark as a universal medicine for healing wounds and aching teeth.[25]

Birch ashes are perfect for potash, carbon black, birch tar and soap. In the old days in Finland, one important product was charcoal. The wintertime

24. Huurre 1995, p. 35; Niemi 2005, pp. 59–61.

25. Kalm 1759, p. 29.

was suitable for burning charcoal, when the ground was covered by snow and rural people had not much other work to do.[26]

In spring, when birch trees begin to green after a long winter, the sap begins to stream into the trunks. Then it is easy to tap the sap. Sap is a healthy drink, or you can make a syrup out of it. People used to make sap beer as well. In the old days, sap was used as a universal medicine. In Germany, distinguished gentlemen drank a glass of sap every day in May against bloating and kidney complications. The ancient name for March in Finnish language was '*mahlakuu*' = 'sap month' since March is a suitable month for tapping sap, when the ground is still a little bit frozen. Cold sunny days around midday are the best moments for tapping sap, when the warmth of the sun has given energy for liquid to rise.[27]

Slash-and-burn cultivation[28]

Slash-and-burn cultivation had a significant impact on Finnish nature for centuries. During a rotation of slash-and-burn cultivation, large areas of forests are in different stages of succession. The duration of a succession depends on the quality of the soil and of the forest. A fertile deciduous forest, after slash-and-burn cultivation, takes from 25 to 40–100 years to mature. Birch and grey alder appeared first to the burned spots. After that, cows, sheep and goats as well as horses grazed in the young forests, until the trees were again mature for felling and burning for cultivation.

26. *Ulrik Rudenschölds...* 1899, p. 62.

27. Niemi 2005, pp.63–64.

28. See Niemi 2005, 46–52.

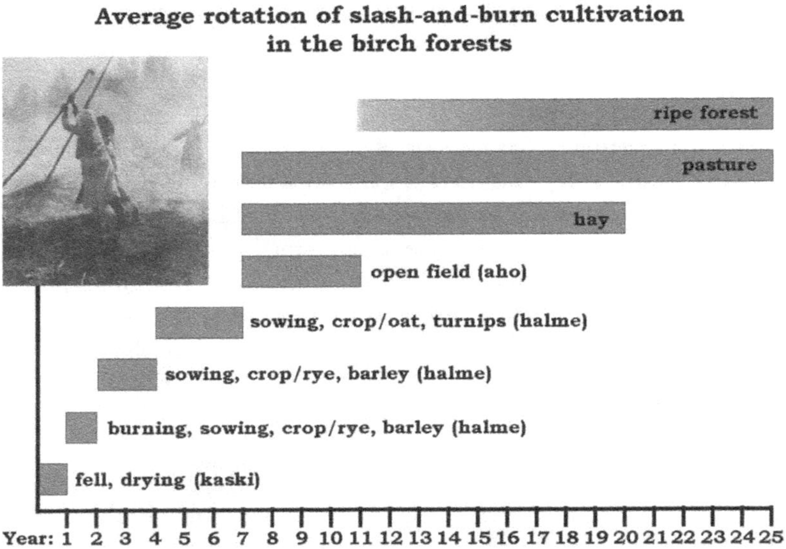

Figure 4. Slash-and-burn rotation. Photograph by the author.

One household needed approximately 1.5 hectares (3.7 acres) for a slash-and-burn cultivation area. For the whole 25 years' rotation, one household needed 25 times 1.5 hectares, so 37.5 hectares (92.66 acres) woodland.

No references are to be found to tell us how old Finns perceived their forests, or whether they had skills that could be called environmental literacy. But we can find indirect references in the huge ancient vocabulary of the Finnish language. For instance, the old Finns used various words for different stages of slash-and-burn cultivation, such as:

'*kaski*' this is the most common slash-and-burn term which can mean:

logging area, area where trees are felled for slash-and-burn cultivation (s.)

as yet unburnt area for s.

logging area, area where deciduous trees are felled for s

burnt area for s.

s.

the land for s., sown area for s.

the crop growing area of s.

'*halme*' the area of growing crops in s.

other sown area, land or area for turnips

'*aho*' the set-aside area of s.

the set-aside area for s. where grass and young forest normally grew

young forest, park, grove, fallow, field, pastureland

Today the Finns still use these words but few know the original meanings anymore.

Edward Sapir has noted:

> ... the 'real world' is to a large extent unconsciously built upon the language habits of the group. ... We see and hear and otherwise experience very largely as we do because the language habits of our community predispose certain choices of interpretation.[29]

Language is in eternal change. The changes in a language happen in interaction with the surrounding society. For instance, when some mode of production, such as slash-and-burn cultivation, stops, people stop using the words related to this mode of production. Klemetti Näkkäläjärvi, who has studied the language of the Sámi people, considers these changes most frightening when they simplify and impoverish the language. They are also definite signs of changed environmental perception.[30] The example of slash-and-burn cultivation vocabulary shows us how the words have survived but their original meanings have faded away. We know still to-day that '*kaski*' means slash-and-burn cultivation, and we know that '*halme*' is a piece of land and '*aho*' is an open field, but the precise meanings of the precise stages of slash-and-burn cultivation are no longer familiar to us. We have also long forgotten the special names for birch species. Today even the difference between a downy birch and a silver birch is unfamiliar to most Finns.

29. Sapir 1958, p. 69.

30. Näkkäläjärvi 2000, 1p. 55.

Figure 5. A silver birch leaf. Photograph by the author.

Figure 6. A downy birch leaf. Photograph by the author.

The environmental literacy of the old slash-and-burn cultivators also included the skill of finding the most suitable forests for the best harvests. Such forests were broad deciduous woodlands rather than dry conifer forests with a soil of moraine or sand. People also knew was the best time to fell the trees, when to burn them, when to sow and when to harvest. Slash-and-burn cultivation benefitted many plant species, such as wild strawberries, daisies, clovers and thistles. Forests were renewed in a natural way. Without slash-and-burn-cultivation, forests would have turned gradually to dark conifer forests.

The contemporary owners of mines and ironworks, state officials and scientists considered slash-and-burn cultivation as severely detrimental to forests. These upper-class people thought that the peasants did not care for their forests at all. They did not understand the rotation of slash-and-cultivation, the benefit of the harvests which the peasants got from burned ground and other useful ways of using the growing forests while waiting for the next burning season to begin the rotation again. These opinions remained predominant since they were those of the ruling class whose word was the truth. So, to this day, public opinion is that the peasants did not respect their forests, while, in fact, forests and fields, their homesteads, were important for their livelihood and it was important for them to hand their homesteads in good shape to future generations. The peasants did not consider their slash-and-burn cultivation as destroying the forests. With the knowledge they inherited from former generations, the peasants knew that the forests would grow again after a certain interval, and they hoped that the next generations would benefit from the forests of their ancestors.

Of course, slash-and-burn cultivation had negative impacts on nature. For example, the spring, when birds nest, was a suitable time for burning the forest. Thus, bird populations declined.

Best firewood in the world[31]

Birch is the best firewood, as everyone in Finland knows. It gives the most heat, it burns nicely and it is easy to torch. In the old days, you could go in your own forest and get as much firewood as you needed. Today you can buy perfect birch firewood almost everywhere in Finland, both in the countryside and in the cities, either on bigger markets or from special sellers. In Central Europe, by the Middle Ages, the firewood business was in the hands of professionals since natural forests had disappeared due to cuttings, pasturing and clearance for fields.

31. See Niemi 2005, pp. 52–54.

In the days when people paid their taxes with different materials, firewood was one of the tax paying instruments to manors, manses, and state buildings such as castles. At times, firewood was among the ten most important Finnish export articles. When Finland was a part of Sweden, about from the year 1100 to the year 1809, lots of firewood was shipped to Stockholm. From 1809 onwards, when Russia conquered Finland, and ruled the country until 1917, when Finland became an independent state, St Petersburg was an important firewood market. After the Saimaa Canal was opened in 1856, and the railroad was built from Helsinki to St Petersburg in 1870, this city became the most important export destination for Finnish firewood.

Before the Second World War, in Finland, more than half of logging in a year was for firewood. Tar burning, and slash-and-burn cultivation accounted for only a tiny part of total forest use. However, within Finnish history research, these other topics have been more prominent than firewood and other household loggings.

In the old times, the worst enemies of birch were people and cattle. Firewood logging, pasturing, suckering, bark taking and chopping had a significant impact on the environment. Short deciduous trees and bushes appeared, and sturdy trees disappeared around villages. In the worst affected areas, severe shortage of wood emerged.

Cultural, scientific and industrial ways to interpret birches in the woods

Cultural birch

The forests have been places where the Finns have sought and found peace of mind. They have also been sources of aesthetic and moral experiences for both the Slavs and other Finno-Ugric people. Birch in particular was a sacred tree. Birch represented purity, goodness, summer and warmth. For the Mordovians, birch was the world tree: sap symbolised the continuity of life and transmigration, branches were symbols of the ancestors and the sphere of heaven.[32]

Fairytales have transported old customs and ideologies through centuries. When patriotism and nationalism started to suffuse societies, nature of the native country became a patriotic symbol. One example from Russia is Sergei Yesenin's poem *Beryoza* (The Birch Tree) which was published in the children's magazine *Mirok* (Small World) in 1914. This poem strengthened warm national

32. https://www.taivaannaula.org/2008/09/26/koivu/ (accessed 6 July 2021).

feelings already present towards the birch tree in Russia.[33] In Finland, a fairytale *Koivu ja tähti* (The Birch and The Star, 1893) by a beloved author, Zacharias Topelius, affected Finns in the same way. The story tells of a boy and his sister, whom the enemy kidnapped during one of the ancient wars between Sweden and Russia. They were brought to Russia where they lived amongst good people. They got enough to eat and were well-dressed; however, when they heard that peace had finally come to their native country after several years, they wanted to return to their father and mother. After a long and difficult journey, they found their home. They identified it from the birch growing in the garden and a star that shone through its branches. Their old mother and father lived still in the cottage. They all were filled with joy when they met each other after so many years. These two words, the birch and the star, became symbols of two important things: the birch for the native country and the star for eternal life after death.[34]

Birch is an essential part of the Finnish national landscape. In the early landscape paintings of the nineteenth century, birch is well represented. One well-known example is Werner Holmberg's *Ihanteellinen maisema* (Ideal Landscape, 1860) which is known also under the name *Suomalainen maisema, vanhoja korkeita koivuja kasvava* (Finnish landscape with high growing old birches). Though Holmberg painted the picture in Norway during his honeymoon, the atmosphere is strongly Finnish with white birch trunks in the foreground and people and cattle somewhere in the distance.[35]

Nature is the base of any culture. Culture and nature are thoroughly intertwined. The artists and poets of the Romantic era turned their gaze to nature. At the turn of the eighteenth and nineteenth centuries, beautiful lake scenes became popular in Finland. A view from a high hill over a region with lakes and forests is the national scenery of Finland. This was a new way to look at the environment. When a farmer of this period looked at this kind of scenery, he saw lakes to be drained and rapids to be cleared to reclaim land for fields.[36]

33. https://fi.ilovevaquero.com/zakon/111525-a-davno-li-bereza-simvol-rossii.html (accessed 6 July 2021).

34. Niemi 2005, p. 128.

35. Ibid., pp. 115–28.

36. Ibid.

Figure 7. An example of the national scenery of Finland. Photograph by the author.

The nineteenth century was a turning point in traditional environmental literacy. New people came into forests such as the above-mentioned artists, but also scientists, surveyors, sawmill owners, industrial bosses and gradually tourists and other recreation seekers.

Scientific birch

One of the first scientific studies on birch trees is *Björckens Egenskaper och Nytta I Den Allmänna Hushålningen* (Properties and Benefits of Birches in Common Households, 1759) by Pehr Kalm, Professor of Economy at the Academy of Turku, with his student Johan Grundberg. Kalm was himself a student of Carl von Linné. A large part of the information in this and other studies was collected by Kalm from common peasants, the Sámi people and from folklore. Most of his information is valid still today.

After Kalm's time, almost a century passed before the scientists found their way to the forests. The first textbook on Finnish forestry was published in 1841. In 1874, a committee found the state of the Finnish forests in devastating decline. The cure would be better knowledge of systematic forestry.

Side by side with the expanding wood processing industry grew forestry as a science. Higher education for foresters began at the Evo Forestry Institute at the end of nineteenth century. The university of Helsinki began educating foresters in 1908.[37]

Birch processing industry[38]

During the nineteenth century, slash-and-burn cultivation was abandoned in the favour of field cultivation. Simultaneously, the need for timber increased considerably within various fields of industry. The traditional way of forestry appreciated forests where all kinds of wood species grew in different stages of their succession. The needs of industry changed these old values. The most suitable forest for certain kinds of industry comprised certain wood species at the same certain age. This new type of forestry changed the appearance of forests totally.

In the first half of the nineteenth century in Germany, some innovative engineers discovered that wood fibres could be used in paper and cardboard production. Before this innovation, the main raw material for pulp and paper was rags. After this crucial turn in the wood processing industry, the value of pine and spruce in Finland grew. The first modern wood processing mills started production during the 1870s. While the value of pine and spruce grew, birch trees were viewed as 'weed trees'. Since birch was not suitable for pulp and paper production, it was despised until the 1960s.

However, the spool and plywood industries appreciated birch. The largest birch forests were in the south-eastern parts of Finland where slash-and-burn cultivation had persevered the longest. The same region was thus suitable for the spool and plywood industry. Fortunately, these industries used different parts of birch, so they did not compete for the same birch resources. Spools were made of thinner trunks while the plywood industry used sturdier birch stocks.

Spool factories[39]

The very first spool factory in Finland was established in Southern Finland, in Mäntsälä. It was built on the banks of Kaukas rapids from which the plant took its name, Kaukas Factory. The founder, Lars Magnus Robert Björkenheim (1835–1878), was a Finnish nobleman who owned nearby Kellokoski manor

37. Niemi 2005, pp. 101–09,
38. Ibid., pp. 143–61.
39. Niemi 2005, pp. 161–63.

and ironworks. Robert Björkenheim died unexpectedly in 1878. Since the birch forests around Kaukas Factory were gradually decreasing, the new owners decided to move production to Lappeenranta, in southeast Finland, closer to better birch wood deposits. The new owners also decided to start pulp and paper production. After some decades, this factory was one of the biggest paper processing units in Finland. Nevertheless, in 1986, Kaukas merged with the future global UPM-Kymmene company.

The spool industry is tightly connected with international economic markets. It depends on consumption of thread and spools, the cloth industry and innovations, economy and trade cycles. Before the Second World War, Finnish spool export headed global statistics. Finnish birch spools were exported, for example, to Great Britain, Russia, France, Germany, Austria, Belgium and Spain. However, in terms of the whole wood processing industry in Finland, the spool industry was only a small part. Kaukas Factory was the first spool factory, and it was also the last one in Finland. When plastic spools replaced wooden ones, Kaukas was closed in 1972. Altogether 23 factories produced spools in Finland in the 100 years during which these were made of birch.

Plywood industry[40]

Plywood is constructed of thin plyboards which are glued together in layers, each with its grain at ninety degrees to the layer beneath. Sufficient supply of raw material is a prerequisite of the plywood industry as are ease of transportation of logs to the factory and of the finished plywood to customers.

The plywood industry was a significant sector of the larger wood processing industry in Europe from the nineteenth century. Most of the plywood was needed for transport cases. For example, tea came to Great Britain from China in plywood boxes. Nordic birch is excellent material for plywood, since it does not give smell or flavour to articles packed in it. The smell of tropical wood species is rather strong and would be easily absorbed by articles transported in cases made from these.

In Europe, the plywood industry used various deciduous tree species, such as beech, oak and birch. In the Baltic countries, birch was used in plywood production in multiple factories. At the beginning of the twentieth century, Russia was the leading manufacturer of birch plywood.

Since Finnish forests had large reservoirs of birch trees it was natural that Finnish entrepreneurs began to produce plywood too. The first plywood factory

40. Niemi 2005, pp. 163–66.

was established in 1893 in southwest Finland, by the Pori-Tampere railroad and near suitable waterways. However, the story of this factory was short: it stopped production after six years because of economic difficulties. Plywood production in Finland started in earnest in 1912, when Wilhelm Schauman, the so-called founder of the Finnish plywood industry, established his factory in Jyväskylä. He had become acquainted with the plywood industry in the United States before starting his own production. He bought machinery from the United States and hired qualified managers from Russia and the Baltic countries. Compared to plywood production in the United States, the Finnish industry used thinner stocks.

After the First World War, the former rulers of the plywood market, Russia and the Baltic countries, were separated from western markets. This period was both advantageous and profitable for the Finnish plywood industry. Finnish plywood was mostly exported to Great Britain in qualities suitable for tea and rubber caskets: these accounted for almost 90 per cent of total production in many years.

In the early days of the Finnish plywood industry, birches were felled both during winter and summer. In winter, logs were piled on sledges driven by horses; in summer, logs were transported on boats. Horse transport was economic over short distances. Rather expensive boat transportation was suitable only in certain kinds of water systems. Gradually, floating became the main way of transporting logs. Since birch is a rather dense and heavy wood species, it gets wet quickly and sinks. Nowadays rail transport is the most common. The logs were stored in piles near the factory. In the summer, logs were easily spoiled and, in the winter, frozen logs cracked easily when soaking in warm water.

When the plywood industry started in the 1910s, people still used birch bark for many household purposes. These old customs threatened to destroy the best birch trunks, since the best bark grow on the straightest and smoothest trunks which also provided the best material for plywood. Gradually, the use of birch bark stopped. The plywood industry also learned to joint, patch and lengthen the plyboards, so they could use all sorts of birches; and the story of the plywood industry has continued until today.

When the plywood industry grew, birch even began to be in short supply. In the first decades of plywood manufacturing, environmental problems did not bother manufacturers. Floating caused problems with many logs lost by sinking. In the soaking pools, harmful extracts were dissolved from the logs into the water. Waste problems were solved by burning, for example bark and the edges of the plywood plates in the steam boilers or selling the waste as fuel to other manufacturers.

Innovative birch

The future of birch as raw material for innovative products looks bright. One innovation is xylitol ($C_5H_7(OH)_5$, D-xylitol). It is made by reducing xylose from xylene which you can get from the leaves of deciduous trees like birch. Xylitol was found by German scientists in 1890s, but it was at the University of Turku in the 1970s that benefits for teeth were discovered. One of the first commercial xylitol products was a chewing gum called Xylitol-Jenkki, launched by a Finnish candy manufacturer in 1975.[41]

A couple of Finnish companies are developing birch bark as an environmentally friendly material for the cosmetic, chemical, textile and pharmaceutical industries. They refine coal powder, bark powder, betulin, suberin and azelaic acid from birch bark. Their aim is to create renewable materials and reduce the use of plastic raw materials.[42]

In 2018, at the presidential reception on Finland's Independence Day of 6 December – a very popular media event that many Finns watch on television – Jenni Haukio, the wife of Finland's president Sauli Niinistö, wore an evening gown made of birch fibres. Since then, the manufacturer has developed the idea of different recycled materials further and transforms used textiles, pulp and even old newspapers into new textile fibres sustainably and without harmful chemicals.[43]

One innovation from the 1930s is the curved armchairs designed by the Finnish architect Alvar Aalto. Today the chairs are globally-known Finnish plywood products. Their triumph began in London in the 1930s when Aalto's designs were exhibited at Fortnum and Mason.

---

41. https://fi.wikipedia.org/wiki/Ksylitoli

42. Sari Möller, Koivun kuoresta jalostetaan muovin korvaajaa kosmetiikkaan. Published 10 Aug. 2021: https://yle.fi/uutiset/3-12053008

43. Ioncell® https://ioncell.fi/ (accessed 13 Sept. 2021).

Figure 8. Alvar Aalto's Paimio-chair, designed by Alvar Aalto for the Paimio Sanatorium in 1932. The seat and back are created from one piece of form-pressed plywood. The fluidity of the design creates a soft, comfortable place to sit. Since wood changes over time, the armrests are formed from a single piece that is then split in half, ensuring that as the chair ages, it remains perfectly balanced. Source: The Finnish Heritage Agency.

Environmental literacy in the Finnish birch forests

Birches in the Finnish forests have encountered various actors. A long time ago, hunters needed birch wood for their bows, skis and other equipment. Later, slash-and-burn cultivators, charcoal burners, wood-hewers and loggers used birch trees for their various needs. These actors had only axes or other simple tools in their hands when they went to the forests. Their impact on nature was minor and there was almost no competition between different users. When lots of wood began to be needed in mines and iron works, and after that in the wood processing industries, problems began to occur.

Before the 1860s not many people other than peasants could read the language of birches in Finnish forests. For peasants, birch was a suitable material for multiple household uses. In the birch forests they also saw suitable areas for slash-and-burn cultivation. When the industrial revolution reached Finland, the socio-economic situation changed. When new paper and pulp mills were established, the value of the forests altered. The ways of environmental literacy changed from those of peasants to those of industrial entrepreneurs. Eventually, the old ways of collecting different raw-materials, slash-and-burn cultivation, grazing, coppicing, suckering, bark taking and so on vanished and the Finnish forests transformed into 'tree fields' for wood processing industry.

When the environment changed, so did economic and political structures and functions: the forests which had been used for rural human purposes became more valuable for the wood processing industry. At the same time, and from another point of view, there arose the need to preserve the forests, and new laws and institutions were established for forestry and conservation. Environmental literacy changed because the ways of thinking about, conceptualising and valuing the surroundings changed. When people abandoned slash-and-burn cultivation they began to read their environment in a new way, and this process also altered their language.

These examples of the changing status of birch trees, how Finns have perceived, used and valued birch over the centuries, are only few fragments of the huge realm of Finnish environmental history. The birch is and has been an important tree for the Finns. They appreciate the silver birch so much that, in 1988, it was voted the National Tree of Finland.

Bibliography

Disinger, John F. and Charles E. Roth. 1992. *Environmental Literacy*. ERIC/CSMEE Digest. http://files.eric.ed.gov/fulltext/ED351201.pdf Accessed 17 Nov. 2013.

Golley, Frank B. 1998. *A Primer for Environmental Literacy.* New Haven and London: Yale University Press.

Hares, Minna, Anu Eskonheimo, Timo Myllyntaus and Olavi Luukkanen. 2006. 'Environmental literacy in interpreting endangered sustainability: Case studies from Thailand and the Sudan'. *Geoforum* 37 (1): 128–44.

Holopainen, Viljo. 1957. 'Metsätalouden edistämistoiminta Suomessa Tapio 1907–1957'. *Silva Fennica* 94. Helsinki: The Finnish Society of Forest Science. Holopainen V. (1957) Promotion of private forestry in Finland, Tapio 1907-1957 (silvafennica.fi) Accessed 7 Nov. 2022.

Hsu, Shih-Jang and Robert E. Roth. 'An assessment of environmental literacy and analysis of predictors of responsible environmental behaviour'. *Environmental Education Research* 3: 229–49.

Huurre, Matti. 1995. *9000 vuotta Suomen esihistoriaa.* 5th edition. Helsinki: Otava.

Häkkinen, Kaisa. 2004. *Nykysuomen etymologinen sanakirja*. Helsinki: WSOY.

Ja Kuinka kauan koivu on ollut Venäjän symboli? [And how long has birch been the symbol of Russia?]: https://fi.ilovevaquero.com/zakon/111525-a-davno-li-bereza-simvol-rossii.html Accessed 6 July 2021.

Kalm, Pehr. 1759. *Oeconomisk beskrifning,* öfwer *Björckens Egenskaper och Nytta i Den Almänna Hushålningen*. Resp. Johan Grundberg. Åbo: Jakob Frenckell.

Laurén, Åke. 1987. *Bondens liv i det gamla Helsinge.* Helsingfors: Söderströms.

Myllyntaus, Timo. 1999. 'Aarniometsistä puupeltoihin: metsät Suomen taloudessa'. In Timo Soikkanen (ed.) *Ympäristöhistorian näkökulmia: piispan apajilta trooppiseen helvettiin.* Turku: Turun yliopiston poliittinen historia.

Niemi, Seija A. 2005. *Suomalaisen metsäluonnon lukemisen historiaa – Ihmisen ja koivun muuttuva suhde Suomessa 1730-luvulta 1930-luvulle.* Licentiate thesis, University of Turku, Department of Finnish history. Turku: Turun yliopisto. Lisensiaatintyo2005Niemi.pdf (utupub.fi)

Niemi, Seija A. 2015. *Koivu. Suomen kansallispuu.* Helsinki: Minerva Kustannus Oy.

Niemi, Seija A. 2018. *A Pioneer of Nordic Conservation: The Environmental Literacy of A.E. Nordenskiöld (1832–1901).* Turku: University of Turku.

Niemi, Seija A. A Pioneer of Nordic Conservation: The Environmental Literacy of A. E. Nordenskiöld (1832–1901): https://www.utupub.fi/bitstream/handle/10024/145775/AnnalesB458Niemi.pdf?sequence=1&isAllowed=y

Näkkäläjärvi, Klemetti. 2000. 'Porosaamelaisten luonnonympäristö'. In Irja Seurujärvi-Kari (ed.) *Beaivvi mánát. Saamelaisten juuret ja nykyaika.* Helsinki: Suomalaisen Kirjallisuuden Seura.

Orr, David W. 1992. *Ecological Literacy. Education and the Transition to a Postmodern World.* New York: State University of New York Press.

Roth, Charles E. 1992. *Environmental Literacy: Its Roots, Evolution and Directions in the 1990s.* Columbus OH: ERIC / CSMEE Publications.

Ruotsala, Helena. 2002. *Muuttuvat palkiset. Elo, työ ja ympäristö Kittilän Kyrön paliskunnassa ja Kuolan Luujärven poronhoitokollektiiveissa vuosina 1930–1995.* Helsinki: Suomen muinaismuistoyhdistys.

Sapir, Edward. 1958 [1929]. 'The status of linguistics as a science'. In D.G. Mandelbaum *Culture, Language and Personality.* Berkeley, CA: University of California Press.

Slotte, Håkan. 1997. 'Lövtäkt – en landskapsdanande verksamhet'. In Lars Östlund (ed.) *Människan och skogen – från naturskog till kulturskog?* Stockholm: Kungl. Skogs- och Lantbruksakademien, Nordiska Museet och Sveriges lantbruksuniversitet.

Svenska Akademiens ordbok: http://g3.spraakdata.gu.se/saob/ Accessed 24 May 2004.

Tomppo, Erkki. 1998. 'Metsävarat vuosina 1989–1994 ja niiden muutokset vuodesta 1951 lähtien'. In Erkki Mälkönen (ed.) *Ympäristönmuutos ja metsien kunto. Metsien terveydentilan tutkimusohjelman loppuraportti.* Vantaa: Metsäntutkimuslaitos.

Ulrik Rudenschölds berättelse om ekonomiska o. a. förhållanden i Finland 1738–1741. 1899. Helsinki: Suomen Historiallinen Seura.

Vallinkoski, Jorma. 1962–1966. *Turun Akatemian väitöskirjat 1642–1828 I.* Helsinki: Helsingin Valtioneuvoston kirjapaino. Turun akatemian väitöskirjat 1642-1828 = Die Dissertationen der alten Universität Turku (Academia Aboënsis) 1642-1828 - I - Doria Accessed 7 Nov. 2022.

Vuorela, Toivo. 1998. *Suomalainen kansankulttuuri.* Porvoo-Helsinki-Juva: WSOY.

Chapter 5

TRASH FOOD? FISH AS FOOD IN FINNISH SOCIETY BETWEEN THE 1870s AND THE 1990s

Matti O. Hannikainen

Introduction

In 1932, a Finnish publication titled *Suomen merikalastus ja jokipyynti* (Finnish Marine Fisheries and River Fishing) listed seventeen fish species as commercially valuable, including Baltic herring (*Clupea harengus membras*), pike (*Esox lucius*) and bream (*Abramis brama*), indicating their culinary value in Finnish society.[1] In the following decades, many fish species became less appreciated, reflecting profound societal changes. These changing values were captured in a recent book *Suomen Kalat* (Fish in Finland) that classified Baltic herring, bream and pike among twenty undervalued and under-used species.[2] In this chapter, we shall therefore analyse why the value of certain fish species has changed and some have been classified 'trash fish', referring to species with little or no value for human consumption, such as silver bream (*Blicca bjoerkna)* and three-spiked spickleneck (*Gasterosteus aculeatus*).

Fishing is the most ancient way of gathering food that remains important globally.[3] Environmental historians, however, have focused on its transformation into an industrial activity that has depleted stocks of the most valuable species, and have thus researched histories of single species, paying only scarce attention to the abundant, albeit economically less important ones.[4] In contrast to the fewer than fifty salt water fish species that are commercially valuable, there are

1. Järvi 1932.
2. Yrjölä et al. 2016, pp. 262–67.
3. Fagan 2018.
4. Greenberg 2000; Svanberg and Locke 2020; Sonck-Rautio 2017.

doi: 10.3197/63824846758018.ch05

some 20,000 fish species altogether, not to mention the freshwater species, suggesting the potential of 'less profitable fish'.[5] This bias concerning fish in history has been addressed by food historians, who have analysed changes in the demand for and consumption of fish. Previous research has shown, for instance, that various species of fish were consumed in huge quantities in England prior to the mid-nineteenth century, when the people began to 'avoid all but a few species and methods of preparation'.[6] In Finnish food histories, however, fish has been taken for granted, with minimal attention paid to changes in valuation and consumption.[7] Thus, this chapter aims to combine environmental history and cultural history of food in analysing perceptions concerning the value of fish in a modernising Nordic country.[8]

Despite the recent animal turn in environmental humanities and environmental history, fish continues to be seen as unfamiliar animals, lacking common emotions with humans.[9] In Finnish culture, for instance, the otherness of fish is captured in numerous proverbs, such as 'like a fish on dry land'.[10] No fish species has been domesticated so as to allow observation of its habits and emotions. Humans and fish are separated by water in addition to which most fish have been caught and farmed for human consumption only; therefore, the human relationship with fish represents perhaps the most one-sided and irreversible human-animal encounter.[11] This chapter also contributes to the ongoing discussion on the animal turn in environmental history, first by moving beyond mammals and secondly by examining the relationship between human and fish in a modern urbanising society, which has attracted only little scholarly attention so far.

In order to analyse the history of human animal-relationships concerning fish in Finnish society, I have identified three discourses – scientific, recreational and culinary. We focus on the scientific and culinary discourses which were crucial in advocating the classification of fish species according to their commercial value and in disseminating societal change, explaining changes in demand in particular. A conceptual analysis of these two discourses is based on critical cross-reading of sources including official documents, such as published

5. Toussant-Samat 2009, pp. 284–85.
6. Freedman 2007.
7. Sillanpää 2006; Kylli 2021.
8. See e.g. Miink 2009.
9. Hoffmann 2002; Brantz 2010.
10. See e.g. Häkkinen 2002; Hakamies 2004.
11. Creager and Jordan 2002; Hoffmann 2002.

reports of parliamentary committees; journal articles on fishing and textbooks on fishing; in addition to cookbooks that serve as pivotal sources to illustrate changes in demand and consumption of various fish species.[12]

Species consumed

Fish has been an intrinsic part of the Finnish diet and is culturally comparable with other Northern European food cultures.[13] There were crucial differences in valuing different fish species, however. The most valued fish species comprised Atlantic salmon (*Salmo salar*), trout (*Salmo trutta*) and whitefish (*Coregonus lavaretus*), referred to as the 'noble stock'.[14] Yet these species were abundant only in a few locations, most notably in their spawning rivers including the Kymi, Kokemäki, Oulu, Ii and Kemi until new hydroelectric power plants ended the spawning in these rivers. The first written records employing the concept 'trash fish' originated from Central and Northern Finland, where species of *Salmonidae* were still abundant in the early twentieth century.[15] However, most Finns did not live close to these rivers, relying on locally abundant species, such as pike, bream, perch (*Perca fluviatilis*), ide (*Leuciscus idus*) and roach (*Rutilus rutilus*). In fact, evidence suggests that, until the 1920s, almost everything caught when fishing was consumed regardless of the number of fish bones or the fat content of the meat, in contrast to a popular perception that only the most valued species were consumed constantly.[16] Hence, regional diversity characterised culinary discourse.

More importantly, seasonality dictated fishing until the early twentieth century, because ice covered the Baltic Sea and the lakes from late November until late April in Southern Finland, whereas in Northern Lapland lakes could still occasionally have ice in June.[17] Whilst winter fishing was practised on a larger scale in certain regions, most notably in the Turku archipelago and the Gulf of Finland with seine netting, fishing mostly concentrated on the spring and autumn spawning seasons that produced huge catches. It was paramount,

---

12. On cookbooks as sources, see e.g. Hannikainen 2022.
13. Hirschfelder 2001, pp. 217–19.
14. Kaski 2019.
15. In the digitised newspapers and periodicals available in the National Archives, 14 references from the period between 1890 and 1920 were found when using the keyword 'roskakala' (literally meaning a trash fish).
16. See e.g. 'Suolaista särpimeksi', Suomen Kuvalehti 1924/25; Rytkönen 1929.
17. Sisävesikalastustoimikunta 1976, p. 4.

however, to cure the catch for winter, a practice which also defined Finnish culinary discourse and the value of most fish species until the 1950s. Thus, the role of fish in the Finnish diet conforms to Paul Freedman's claim that 'food reflects the environment of a society but is not completely determined by it'.[18]

The curing method was selected according to the fish species in question, with different methods for curing each species. The preferred method for curing fish was salting, although drying and even fermenting were employed in the nineteenth century. According to oral histories collected from the Keuruu region in Central Finland, those species with more fat content, such as perch and roach were fermented, whereas drier species, like pike and burbot (*Lota lota*), were dried.[19] The first notable change in Finnish food culture was the gradual replacement of fermentation by salting that began during the seventeenth century when cheaper salt was imported in larger quantities. However, it was not until the second half of the nineteenth century that salting finally overtook fermentation in more remote regions of the then Grand Duchy.[20] This change was underlined by the fact that fermenting fish was only rarely mentioned as a curing method in cookbooks published after the 1890s. In contrast, drying survived in cookbooks until the 1950s, but as a curing method it was widely used only during the war years in the twentieth century.[21] In fact, almost all cookbooks published between the 1890s and the 1970s provided their readers with either instructions for salting fish or recipes for cooking salted fish, often both. One of the last recipes for curing a whole fish by salting was published in the early 1990s.[22]

Table 1. Population of Finland 1870–2020.

Year	1870	1900	1930	1970	2000	2020
Population	1,786,800	2,655,900	3,462,700	4,598,336	5,181,115	5,553,793

Source: Suomen tilastollinen vuosikirja 2022, 257, taulukko 12.9: https://www.doria.fi/bitstream/handle/10024/186220/yyti_stv_202200_2022_25871_net.pdf (Accessed 27 Dec. 2022).

18. Freedman 2007, p. 8.
19. Lappi (ed.) 1996, pp. 87–94.
20. Talve 1961, p. 20; Sillanpää 1999, p. 26.
21. See e.g. Valtion kotitaloustoimikunta 1918; see also Koskimies and Somersalo 1932; Walli 1933; Arppe 1940, p. 331; H.A 1942.
22. Suomalainen and Muurinen 1991, p. 9.

Continuous population growth, as shown in Table 1, coupled with accelerating urbanisation and industrialisation had a profound impact on the consumption of fish and also on culinary discourse after the 1870s. New cookbooks were published because of urbanisation; in a rural society, the art of cooking was based on oral tradition instead of written instructions. The aim of the first cookbooks published in Finnish between 1890 and 1920 was to provide recipes for young working class women to prepare healthy and nutritious food for themselves and their families. The first cookbooks written in Finnish had recipes for the cheaper, hence less valued, species, such as roach, rudd (*Scardinius erythrophthalmus*), perch and bleak (*Alburnus alburnus*) instead of Atlantic salmon and whitefish.[23] In contrast, the more valuable species were mentioned in cookbooks published for wealthier middle-classes, who could afford Atlantic salmon, eel (*Anguilla anguilla*) and pike-perch (*Sander lucioperca*).[24] *Kotiruoka* (Home Cooked Food), published in 1908, provided recipes to a wider readership than the aforementioned cookbooks, which was evident in the selection of fish species ranging from the most expensive to the cheapest species, such as Baltic herring, although the range was surprisingly limited, consisting only of perch, pike, ruffe (*Gymnocephalus cernua*), pike-perch, burbot, vendace (*Coregonus albula*) and lamprey (*Lampetra fluviatilis*).[25] Hence, the consumption of fish remained unselective reflecting the low income level of most Finns and the importance of subsistence fishing until the early 1920s.

Culinary discourse began to emphasise fresh fish instead of salted in the inter-war period. The increasing commercialisation of fisheries and the improvement of transportation between fisheries and consumers improved the availability of fresh fish in urbanising communities. This change was captured in the increased list of species the fourteenth revised edition of *Kotiruoka* (Home Cooked Food) published in 1938, suggesting that the major cookbooks reacted to changes instead of advocating novelties. The new species listed were sprat (*Sprattus sprattus*), smelt (*Osmerus eperlanus*), bream, whitefish, ide and cod (*Gadus morhua*). The revised edition, moreover, contained more recipes for fresh than salted Baltic herring, indicating a change in consumer preference given the fact that Baltic herring was not only the most important commercial fish, but the most preferred salted fish.[26] Reflecting preference for fresh fish was the introduction of filleting fish. Instead of cooking whole fish, a new

23. Friberg 1900; Forstén 1902.
24. See e.g. Kunnas 1914.
25. Reinilä et al. 1908.
26. Reinilä-Hellman et al. 1938.

cookbook *Keittotaito* (Art of Cooking), published in 1932, provided the first instructions how to fillet a fish with four recipes for fillet dishes mainly using pike.[27] Underlining the novelty of filleting in Finnish cooking, the method was introduced a few years later in *Kotiliesi*, an influential women's journal published since 1922, disseminating it for much wider audience.[28] Yet filleting appeared an opportunity to increase the consumption of fish by preparing any fish in the desired boneless way. While preferences for cooking fish were changing, the culinary discourse continued to encourage Finns to consume fish species unselectively.

Species classified

Finland was one of the last European countries where subsistence fishing remained more important than commercial fishing, in contrast to the UK or the German kingdoms (from 1871 the German Empire).[29] Limited commercial fishing had concentrated on the coastal regions close to the largest towns, such as Pietari (St Petersburg), Viipuri (Viborg), Turku (Åbo), Stockholm and Helsinki (Helsingfors). The dominant role of subsistence fishing was one of the reasons to modernise fishing through a new scientific discourse. The advent of this discourse was also linked with the famine that Finnish society experienced between 1866 and 1868 due to failed crops, leading to starvation of at least 130,000 people.[30] Whilst the famine initiated the modernisation of agriculture, it also accelerated the discussion on modernising fishing.

Despite the appointment of an Inspector of Fishing in 1861, it was not until the late 1880s that the new discourse on fishing was formulated. The key person was Oscar Nordqvist (1858–1925), who was appointed the Inspector of Fishing in 1889. Following a seven-week study tour in autumn 1890, visiting fisheries in Germany, England, Scotland and Sweden, Nordqvist realised that fishing in the Grand Duchy should be modernised to feed the nation more efficiently. In Prussia, for instance, the annual yield from the lakes was estimated at between 16 and 45 kilos per hectare in addition to carp ponds that yielded nearly eighty kilos per hectare, in contrast to Finnish lakes that yielded approximately five kilos. Thus, the tour had a profound impact on the emerging scientific discourse of fishing. The new discourse aimed at transforming

27. Koskimies and Somersalo 1932.

28. Tennberg 1936.

29. Walton 2000.

30. Jutikkala 2003.

subsistence fishing into a commercial industry with modern equipment and processing facilities. More importantly, the new discourse sought to classify all fish species according to their commercial value to increase the fishers' income.[31] In addition, the discourse aimed at controlling, if not preventing, fishing during spawning seasons in order to safeguard reproduction – with little success, because spawning seasons provided the largest catches.[32]

The advent of this discourse coincided with a paradigm change with respect to human-animal relationships. The new paradigm classified all animals (mammals, birds and fish) according to their value from a human perspective thus advocating complete human mastery over nature.[33] The new scientific discourse on fishing followed this new paradigm extending the human control to the species living beneath the waves.[34] More importantly, the scientific discourse on fishing followed a commercial logic separating it from the other human-animal discourses and making it the most controversial and complex. No fish species posed a threat to humans, farm animals or crops – they were only harmful to other species of fish, either in competing for the same food or in feeding on other fish perceived as more valuable by humans.[35] Further underlining the human mastery over fish, all fish species were protected from predators, such as otters (*Lutra lutra*), black-throated divers (*Gavia arctica*) and Saimaa ringed seals (*Pusa hispida saimensis*), and hunting of these species was encouraged by bounties, a policy lasting from the 1880s until the mid-1920s.[36]

The new discourse was disseminated in fishing manuals and textbooks on fish. These publications were part of the building project of an emerging independent nation that required research on the species 'naturalised' within Finnish boundaries.[37] The impact of the new discourse was visible in the two editions of *Suomen selkärankaiset* (*Finnish Vertebrates*). In the first edition published in 1882 all fish were classified as edible, although a few were described as worthless because they were either pelagic or bottom feeders and therefore rarely caught, limiting their consumption and commercial value. The second edition published in 1909 was rewritten to comply with the new scientific discourse. Accordingly, all fish species were classified into either commercially

31. Nordqvist 1891; see also Gottberg 1913.
32. Kalastuskomitea 1911, p. 17; *Metsästys ja kalastus* 1912.
33. Ilvesviita 1995.
34. See e.g. Räsänen 2021, p. 280.
35. Vuorisalo and Oksanen 2021.
36. Järvi 1941; Vuorisalo and Oksanen 2021.
37. Räsänen 2021, p. 277.

valuable, regionally valuable or worthless. Yet this later edition did not employ the sinister definition characteristic of the scientific discourse.[38] In 1902, Oscar Nordqvist published a textbook that classified all fish species spawning in Finland according to their commercial value in the three categories: commercially valuable, less valuable and worthless trash fish. For instance, most species from the family *Cyprinidae* were classified worthless with silver bream listed as a trash fish, because its 'meat is loose, thin and very bony, which is the reason, why it must be considered trash among the fish and must be, if possible, exterminated and replaced with the bream'. In contrast to the earlier textbooks, Nordqvist openly proposed the extermination of trash fish that were considered to compete for food with more valuable species, such as bream in this case.[39]

The new scientific discourse notwithstanding, the annual catch grew only gradually prior to 1914. Fishing intensified to feed the growing population (Table 1), although its commercialisation was hampered by constant problems with seasonality, transportation and lack of investment. Between 1901 and 1913, the annual catch increased only from some fourteen million kilos to some twenty million.[40] Yet these numbers exclude most freshwater catch, because in the coastal regions and in the vicinity of the major spawning rivers fishing was more commercially organised and therefore more reliably recorded. The scientific discourse affected species classification in official statistics. In 1901, for instance, the official statistics comprised six categories: the first consisted of *Salmonidae*; the second whitefish and grayling (*Thymallus thymallus*); the third Baltic herring; the fourth vendace and sprat; the fifth smelt; and the last all other species grouped together as 'other species' (lit. '*muut kalat*').[41] This categorisation was replaced 1931 by a new classification that covered both freshwater and marine fishing in separate categories. These accommodated mostly individual species, reflecting the impact of commercialisation of fishing. For marine catch, the first category comprised *Salmonidae*; but the the

38. Mela and Kivirikko 1909.

39. Nordqvist 1902.

40. Suomen tilastollinen vuosikirja (STV) 1903: https://www.doria.fi/bitstream/handle/10024/67177/stv_1903.pdf (accessed 13 May 2021); Luonnonvarakeskus, Tilastotietokanta, Suomen kalastuksen saaliit (1000 kg) 1980–: http://statdb.luke.fi/PXWeb/pxweb/fi/LUKE/LUKE__06%20Kala%20ja%20riista__02%20Rakenne%20ja%20tuotanto__08%20Kalastus%20yhteensa/03_Kokonaiskalansaalis.px/ (accessed 30 June 2021).

41. Suomen tilastollinen vuosikirja 1903, 71. Kalan- ja hylkeenpyynti vuosina 18771901: https://www.doria.fi/bitstream/handle/10024/67177/stv_1903.pdf (Accessed 28 Dec. 2022).

second consisted of whitefish; the third Baltic herring; the fourth sprat, the fifth smelt, the sixth pike, which was introduced as a new species; and all other species grouped together as the last category. Freshwater catch was classified similarly. The first category consisted of *Salmonidae*; the second grayling, the third whitefish, the fourth vendace; the fifth pike and the sixth pike-perch as new species; and the rest, such as bream, ide and burbot grouped as other species.[42] New categories reflected increasing interest in concentrating fishing on the commercially valuable species following the classification advocated by the scientific discourse. Despite attempts to intensify fishing and to increase its commercial viability, many species classified as less valuable, even trash, in the scientific discourse remained consumed by common people, given the importance of subsistence fishing in the more remote countryside.[43]

The turbulent decade 1910–20, which witnessed Finland gaining independence, the new nation surviving a fierce civil war and finally signing a peace treaty with Soviet-Russia in October 1920, left commercial fishing in precarious situation. Above all, the lucrative trade to the imperial metropolis, St Petersburg, ended leaving numerous fishing communities in South-Eastern Finland looking for new markets for Baltic herring, ruffe and smelt, without much success during the interwar years. The chronic lack of investment in processing the catch and providing speedier transportation to the growing urban communities in Southern Finland particularly affected the consumption of fish.

During these years, the role of the Suomen Kalastusyhdistys (Finnish Fisheries Association, 1891) was important, albeit controversial. Its members promoted scientific discourse and the modernisation of fishing into a commercial industry, thus classifying fish species according to their commercial value. The association, however, promoted unselective consumption of fish in order to boost consumption, thus supporting fishers by publishing numerous cookbooks specialising in fish. The first such cookbook was titled *Yksinkertaisia Kalaruokia* (Simple Fishfood) published in 1918, which was also one of the first cookbooks specialising in fish written in Finnish. Reflecting the contemporary art of cooking, its recipes were divided into dried, salted and fresh fish with recipes for different fish species.[44] Given the problem with Baltic herring, the association attempted to increase its domestic consumption by labelling it the most important fish commercially in its journal and by publishing separate

---

42. Suomen tilastollinen vuosikirja 1932, 86. Kalastus vuonna 1931 läänittäin: https://www.doria.fi/bitstream/handle/10024/69225/stv_1932.pdf (Accessed 28 Dec. 2022).
43. See, for instance, Rytkönen 1929; Artukka 1925.
44. Artukka 1918.

cookbooks in the 1920s and 1930s.[45] Gradually growing prosperity increased the demand for fresh fish instead of salted fish, causing the association to attempt improvement in quality of salting, with limited success, however.[46] In 1934, the association published another cookbook titled *Kalaruokia* (Cooking Fish) written by Kerttu Olsoni, in order to increase the consumption of domestic fish species. Olsoni argued for unselective consumption of fish even there were no recipes for the trash fish, such as three- and ten-spiked spickleneck, silver bream and blue bream, underlining the controversial character of the culinary discourse.[47]

The 1920s witnessed some of the fiercest writings on trash fish. These were fuelled by growing concern over the sustainability of the most valuable species, most notably Atlantic salmon, due to the construction of hydroelectric dams and the increasing pollution from industry. In June 1923, the limnologist Heikki Järnefelt, who served later as a professor at the University of Helsinki, wrote in an article published in a regional newspaper *Turun Sanomat* that 'we should wage a ruthless war aiming to remorselessly exterminate all those fish we cannot use commercially' referring to roach, rudd, ruffe and silver bream. Järnefelt ended the article with the sinister words 'above all, annihilate trash fish!'[48] Despite this kind of malicious rhetoric, the members of Suomen Kalastusyhdistys recognised the difficulties of transforming the discourse into policy.[49] Above all, Finnish law safeguarded private fishing rights of landowners that effectively prohibited commercial exploitation of an abundant natural resource, including extermination of species. Surprisingly, nature conservation discourse that was codified into a new law in 1923 omitted fish, reflecting common attitudes in Finnish society.[50] Whilst the new law discontinued paying bounties for fish-eating predators, it neither preserved any fish species nor criticised commercial fishing.[51] Thus, the discourse on annihilating trash fish could continue.

The complexity of the scientific discourse was illustrated in the 1930s. Two seminal textbooks on fishing were published, which followed classification of fish species according to their commercial value. *Suomen merikalastus*

45. See e.g. Reinilä-Hellman 1925; Nenonen 1938.
46. Haapala 1926.
47. Olsoni 1934.
48. Järnefelt 1923.
49. See e.g. Hellevaara 1927.
50. Ilvesviita 2005, pp. 344–45.
51. Pohja-Mykrä and Mykrä 2007, pp. 182–83.

ja jokipyynti (Finnish Marine Fisheries and River Fishing, 1932) and *Suomen Kalat* (Fish in Finland, 1934) both employed the three-tier classification for each species of fish according to the structure and taste of their meat, thus defining commercial value in accordance with scientific discourse. Whilst both books classified numerous species worthless, there were only a few trash fish, such as silver bream which 'should be annihilated' – the vicious, albeit logical, outcome of the scientific discourse. Yet both books underlined the selectiveness of the concept. Roach, for instance, was classified a less-valuable species given its numerous bones, watery meat and small size, but it was economically important in certain regions in addition to which it was vital for pike, because pike would transform the less-valuable meat of roach into more valuable flesh that was always in demand for a better price.[52]

The Second World War and wartime food rationing revealed how the consumption of fish had changed during the preceding decades. Both demand and supply had concentrated on the few commercially valuable species indicating the devaluation of many species, most notably roach, that had been valued previously. The closure of the Gulf of Finland, in particular, to all fishing increased unselective consumption of fish with drying once again being practised.[53] Despite the importance of cod in commercial fishing globally, most Finns were unfamiliar with cod, indicating the limited reach of commercial fishing even close to the major urban centres such as Helsinki, in addition to the preference for salted fish instead of dried. General reluctance to cook cod caused surprisingly bitter comments in wartime cookbooks given its availability when numerous fish species were rationed.[54] Some less valuable species, such as roach, ruffe and rudd, also enjoyed a brief consumption peak given the scarcity of options, but their consumption fell as soon as fishing returned to normal conditions by the late 1940s.[55] The government authorised unregulated fishing that lasted from 1941 until 1948 with a few exceptions, but fishing professionals were concerned that unregulated fishing might generate overfishing that would deplete stocks of the most valuable species. Subsequently, new fishing legislation enacted in 1951 reasserted previous restrictions.[56] Yet we know too little about the impact of wartime on the valuation of fish, given the numerous changes occurring in the following decades.

52. Valle 1934.

53. Kallioniemi 2013, pp. 10–11.

54. Olsonen 1940; Nissinen and Somersalo 1943.

55. Pukkila 2008.

56. Kalataloudellinen komitea 1951, p. 18; also Lappalainen 1995, pp. 66–69.

New species

In the post-war years fisheries became more efficient and in the early 1950s Finland was self-sufficient in fish despite a constantly growing population.[57] In particular, the mechanisation of marine fisheries such as the introduction of trawl and specialised techniques for Atlantic salmon, for example, led to an increase in the total catch, although the number of professional fishermen declined from the 1940s to the 1960s.[58] While the annual catch was nearly sixty million kilos in 1953, fishing intensified with the annual catch reaching 123 million kilos in 1980 and later increasing to a staggering 156 million kilos.[59] Most of the annual catch, however, comprised Baltic herring – its share being 31 million kilos in 1953 and growing to some 90 million kilos in 1980.[60] While Baltic herring remained the most commercially valuable species, given the quantity of the catch, it was fed to farm and fur animals and to farmed fish in huge quantities from the 1960s. Unfortunately, the electrification of the nation required constructing hydroelectric power plants in numerous rivers, which, in addition to the increasing pollution, depleted the stock of many anadromous species, most notably family *Salmonidae*. Paradoxically, whilst fishing became more efficient, the stocks of the commercially valuable species decreased, reducing the overall value of the catch and indicating a need to revise the scientific discourse.[61] (Appendix 1)

Simultaneously, Finnish society transformed from an agrarian society into a modern urban society. The urbanising society moreover industrialised and electrified; hence, in the numerous new homes, modern electric novelties, such as stoves, refrigerators and freezers, proliferated from the mid-1950s.[62] This reduced the previous need to cure fish for the winter period by salting. Consequently, fresh fish replaced salted fish, marking one of the greatest changes in Finnish food history. The collapse in consumption of salted fish diminished the consumption of Baltic herring, causing the parliament to appoint

57. Suomen FAO-toimikunta 1952.

58. Kalatalouskomitea 1965, pp. 4–5.

59. Kalan markkinointitoimikunta 1974, p. 72; Luonnonvarakeskus, Tilastotietokanta, Suomen kalastuksen saaliit (1000 kg) 1980–: http://statdb.luke.fi/PXWeb/pxweb/fi/LUKE/LUKE__06%20Kala%20ja%20riista__02%20Rakenne%20ja%20tuotanto__08%20Kalastus%20yhteensa/03_Kokonaiskalansaalis.px/ (Accessed 30 June 2021).

60. Koli 1990, p. 13.

61. See e.g. Muuttuvien vesistöjen kalatalouden hoitotoimikunta 1967, pp. 1–3.

62. Knuuttila 2013.

two separate committees to investigate the case. Yet salted herring remained a crucial ingredient in Finnish diet for many until the 1960s, whereas fresh herring remained popular until the 1990s.[63] There was also external pressure. The imports of frozen fish mainly from Norway intensified due to Norwegian export policy. Previously, imports had consisted of various dried fish.[64] Now imports skyrocketed from 0.8 to 6.8 million kilos between 1958 and 1973 with frozen boneless fillets becoming a favourite of city dwellers.[65] As more women found employment as part of the post-war economic boom and new social welfare induced policy, frozen fillets were preferred for their ease of cooking and eating. More importantly, the depletion of Atlantic salmon and increasing imports of frozen fish encouraged forays into farming foreign fish species.[66] In the 1960s, the farming of rainbow trout (*Oncorhynchus mykiss*) was successfully reintroduced, and it boomed, with production reaching some 5.4 million kilos by the mid-1980s. The reasons for the popularity and growing consumption of rainbow trout were its constant supply and quality at an affordable price compared to Atlantic salmon.[67] One of the significant changes in Finnish environmental history has been the introduction and popularity of farmed fish since the 1970s.

These changes prompted parliament to set up committees to investigate the state of fishing and fisheries in the 1950s and 1960s; these reports were controversial indicating a mounting pressure to revise the scientific discourse. In 1951, a committee investigating fishing of Baltic herring acknowledged the role species classified trash fish, such as roach, rudd, blue bream and silver bream, could play as a food reserve in future.[68] However, a few years later, another committee analysing freshwater fishing recommended the annihilation of the trash fish to revive stocks of the commercially valuable freshwater species, most notably lake salmon (*Salmo salar m. Sebago*) and lake trout (*Salmo trutta m. lacustris*).[69] In 1967, a committee reported that fishing had changed profoundly in northern Finland as a result of the construction of dams and the increasing pollution by industries, so that 'the most valuable anadromous fish species', such as Atlantic salmon, trout, whitefish and lamprey were on the brink of

63. e.g. Sillanpää 1999, p. 205.
64. Levander 1924.
65. Kalan markkinointitoimikunta 1974, p. 144.
66. See e.g. Peltoniemi 1984.
67. Järvi 1941; Peltoniemi 1984; Kalateollisuustoimikunta 1985; Laitinen (ed.) 2014.
68. Kalataloudellinen komitea 1951.
69. Järvikalastuskomitea 1958.

extinction. In contrast, the number of the less valuable species was growing, forcing the fisher to fish species that were 'not commonly used for human consumption'. While recognising the precarious situation, the committee opted to propose a new two-tier classification, dividing fish into commercially valuable species and less valuable species that were unsuited as food.[70] Despite increasing environmental awareness and nature conservation, the scientific discourse on fishing continued to consider certain fish species expendable.

The declining demand for native fish caused the parliament to appoint new committees in the 1970s. The first committee delivered its report in 1974, attributing the decline in fish consumption to the exponentially growing consumption of imported frozen fish, whereas supply of domestic fish was hampered by poor marketing and constant quality problems. The committee noted moreover that consumption had become more selective, with the tastiest species being in constant demand regardless of price, which would further deplete their stocks.[71] Two years later, the second committee delivered a report that changed the scientific discourse. Acknowledging that the stocks of the most valuable species could no longer sustain effective fishing, the committee replaced the concept trash fish by that of less valuable species in order to boost their marketing and consumption in contrast to exterminating them. The committee moreover recommended more intense product development and fish farming.[72]

These changes were eventually captured in numerous cookbooks with more recipes for fresh than salted fish. *Kotikokki keittää* (Home Chef Cooking) published in 1956 embodied the transformation in culinary discourse. Fresh fish and filleting of fish were the basics for preparing a fish dish. The book moreover provided a recipe for an 'Icelandic fillet', one of the first appearances of frozen imported fish in Finnish cookbooks.[73] Both the revised edition of *Keittotaito* (Art of Cooking) and the completely revised 23rd edition of *Kotiruoka* reflected these changes. More importantly, the new edition of *Kotiruoka* (Home Cooked Food) featured a list of most suitable fish species for cooking chosen based on their perceived characteristics, such as taste, smell, fat content and the number of bones. For example, the smelt had a 'revolting smell', and herring was a 'cheap, fatty, and valuable fish in nutritional value', while walleye was 'a fine-tasting fish with few bones'. The fact that the fish list had identification pictures suggested that readers were no longer assumed to be able to recognise

70. Muuttuvien vesistöjen kalatalouden hoitotoimikunta 1967, pp. 1–3.

71. Kalan markkinointitoimikunta 1974, pp. 56–57, 144.

72. Sisävesikalastustoimikunta 1976, p. 48.

73. Wartiainen and Tolvanen 1956.

fish species.[74] The listing also indicated the adaptation of scientific discourse into culinary discourse.

The scientific discourse changed nominally after the 1970s. While textbooks published after 1970 rarely employed the concept of trash fish in classifying species, the concept continued as a sideline. For instance, in a textbook *Suomen Eläimet 3* (Finnish Fauna 3), that dealt with fish and reptiles found in Finland, the concept of less valuable species was employed to classify species like roach and silver bream. *Otavan kalakirja* (Otava's Fishbook, 2001) employed the concept of less valuable species similarly. However, in both books, the concept of trash fish was still employed in captions.[75] Textbooks thus continued to classify fish species according to their commercial value and taste, though in the new century they aimed at encouraging Finns to consume more domestic fish while still acknowledging that some fish were trash.

Notwithstanding increasing environmental awareness that had succeeded in protecting numerous species, fish remained exploitable species in the 1970s. Apart from anadromous species, fish species classified as less valuable could be netted *en masse* from lakes that suffered from oxygen deficiency. Despite the fact that, in most cases, the ultimate culprit for polluted water was human activity due to massive use of fertilisers, for example, there was hardly any criticism of mass removal of less valuable fish which were simply left to rot on the fields nearby as trash. The new environmental policy advocated improvement of lakes as well as the Baltic by mass removal of the less valuable species such as roach, rudd and silver bream, echoing the idea of exterminating trash fish that had been once part of a scientific discourse. The tragic case of mass removal of trash fish exemplifies the ultimate paradox concerning fish as part of human-animal relationships in Finnish society.

Paradoxically, when trash fish disappeared from scientific discourse, it surfaced in culinary discourse. One of the first cookbooks that exemplified this transition was *Maukasta kalasta* (Delicious Fish Dishes, 1973). It was the first cookbook to employ 'trash fish' and also one of the first to offer a recipe for rainbow trout.[76] Two years earlier, the Finnish Fisheries Association had organised a competition for the best domestic cookery recipe for fish. The recipes were divided into three categories according to the value of the species: vendace and Baltic herring; other valuable species including pike, whitefish, Atlantic salmon, perch, walleye and bream; and less valuable species. Out of

74. Lehtinen and Salme 1962; Tennberg and Rautiainen 1960, pp. 185–87.

75. Koli (ed.) 1984; Koli 200, p. 64.

76. Kolmonen and Vanamo 1973.

nearly 600 recipes received, 31 were published in a cooking booklet, including five recipes serving roach as the main ingredient.[77] After the 1970s, cookbooks reflected the increasing affluence of Finnish society coupled with imports of frozen fish and the introduction of farmed fish. The change in Finnish fish consumption was aptly spelled out by Lars Johansson who, in 1987, lamented the increasing imports of fish from abroad while Finns consumed less domestic species, most notably Baltic herring.[78] In fact, the consumption of domestic fish decreased between 1980 and 2000 while that of imported and farmed fish grew.[79] By 1999, the overall consumption of fish was 12.1 kilos per capita of which the proportion of imported fish was six kilos.[80]

The growing popularity of farmed fish and imported fish was evident in cookbooks published in the 1980s and 1990s. The culinary discourse diversified and specialised, reflecting the broader range of species available and the more selective consumption of fish. The number of imported species grew from a mere four mentioned in the late 1970s to nine by the late 1990s.[81] General cookbooks and special cookbooks on fish preferred the commercially valuable species including imported and farmed species, whereas there were only limited number of cookbooks that dealt with less valuable species, often classified by the former as trash fish. More importantly, the number of recipes for rainbow trout outnumbered those for Baltic herring.[82] A committee investigating the consumption of domestic fish that published its report in 1990 noted that the valuable species remained in more demand than the less valuable species despite the fact that fishing of Baltic herring, Atlantic salmon and trout, cod, vendace and whitefish had reached their maximum sustainable point. These species remained popular, despite their high price, variable availability and quality compared with imported fish. The committee referred to the depreciation of less valuable species which had 'no significant use for human consumption'.[83] This said, a recent cookbook promoted recipes for species such as roach, ruffe, blue bream and ide, which had become less valuable during the past century.[84]

77. Suomen kalastusyhdistys 1972.
78. Johansson 1987, p. 81.
79. Partanen 2017, p. 7.
80. Luonnonvarakeskus, Kalan kulutus 2018: https://stat.luke.fi/kalan-kulutus-2018_fi (Accessed 12 Jan. 2022).
81. Nordlund 1978; Lindstedt 1997.
82. See e.g. Liimatainen 2003.
83. Kotimaisen kalan toimikunta 1990, p. 39.
84. Vetikko et al. 2009.

Fish as food

To conclude, in this chapter we have analysed how the value of various fish species has evolved in Finnish society, thus combining two discourses – scientific and culinary – as well as cultural history and environmental history. More importantly, this chapter is pioneering in discussing fish as part of human-animal relationships in a modern society. The relationship between Finnish people (excluding the Sámi) and fish was dominated by a human perspective allowing almost total human exploitation of fish with little, if any consideration, of their agency. This mentality continued throughout the period studied. First came the scientific discourse initiated in the late nineteenth century as part of the modern concern with controlling and exploiting nature for human benefit, which promoted classification of all fish species into three categories – commercially valuable, less valuable and trash – according to a commercial logic. This followed the demand for different fish species, favouring those species with tastier and more nutritious meat and fewer bones. While the scientific discourse aimed at increasing the maximum income of fishers, it denied all fish their agency – in contrast with subsistence fishing not to mention modern recreational fishing. This said, fishing remained a marginal industry and its economic importance was minimal in Finland throughout the twentieth century. Despite the aggressive rhetoric of the scientific discourse, it remained relatively unsuccessful in turning its aims into effective policies.

The concept trash fish contradicts many perspectives in current discussions of human-animal relationships, which have focused mostly on household pets, domesticated animals and larger beasts each familiar to humans as mammals with shared environments. In contrast, fish continue to represent perceived otherness because of living in a different environment, despite the fact that fish shares numerous behavioural patterns with other animals – above all, they do feel pain. Yet, the modern relationship between fish and Finns was driven by commerce and its historical analysis has demonstrated that this perceived otherness allowed the scientific discourse on their use to repeatedly suggest exterminating species considered trash until the mid-1970s. Nature conservation and environmentalism, which affected other human-animal discourses, hunting in particular, developed and gained popularity as Finnish society urbanised, but fish and fishing were rarely touched in the emerging environmental discourse. In fact, the mass removal of fish carried out after the 1970s to improve poor water quality caused by human activity stands out as one of the ultimate cases of animal exploitation in Finnish environmental history, a topic which requires more research.

Our analyses on which fish species have been valued and devalued provides us with new perspectives on what Finnish people have eaten and valued as food and how modernising society has affected something as mundane as cooking. In contrast to the scientific discourse, the culinary discourse shows that Finns consumed almost every fish species caught until the early twentieth century. Whilst culinary discourse was characterised by unselective consumption, seasonality and region, not to mention the dominance of local subsistence fishing, societal change and the general modernisation of society, such as the introduction of new means of transportation and electrification and the commercialisation of fishing, transformed previously unselective local subsistence fishing into selective industrial fishing dictated by affluent consumerism.

The post-war decades therefore witnessed two profound changes in culinary discourse on fish. First the consumption of cured, mainly salted, fish was replaced by demand for fresh, preferably boneless, fish fillets, marking the greatest change in Finnish food culture. Subsequently, the valuation of and demand for species preferred as salted, such as smelt, roach, Baltic herring and ide, waned. Secondly, the introduction fish farming with rainbow trout and the subsequent boom in demand for farmed fish coupled with demand for imported fish affected the demand for and consumption of local domestic fish species excluding, of course, 'the noble stock'. Whilst this has the positive impact of broadening the range of fish species available, the consumption and valuation of numerous domestic species, such as pike and bream, have decreased.

The concept of trash fish thus provides us with new perspectives on traditional cooking methods and preferences in Finnish society, revealing how unselective consumption partially adapted to, partially dictated by environment, changed to selectiveness. Farmed and imported fish species, however, connect Finnish consumers with global commercial fishing and its numerous environmental problems. Recent research suggests that, by the 2030s, most commercially exploited marine fish stocks will be depleted beyond commercial use.[85] Perhaps a historical analysis may provoke Finnish people to understand the importance of domestic, currently less valued, fish species and change their consumption to be more varied and local, and thus better.

Acknowledgement

This research was funded by Maj & Tor Nessling Foundation.

---

85. See https://wwf.panda.org/discover/our_focus/freshwater_practice/the_world_s_forgotten_fishes/ (Accessed 23 Feb. 2021).

Bibliography

Arppe, H. 1940. 'Kalojen kuivaaminen talveksi'. *Kotiliesi* 13: 331.

Artukka, A. 1918. *Yksinkertaisia kalaruokia*. Helsinki: Suomen kalastusyhdistys.

Artukka, A. 1925. 'Kevään kalansaaliin varalta – muutamia säilytystapoja ja maukkaita ruokalajeja'. *Kotiliesi* 10: 274–275.

Bogue, M.B. 2000. *Fishing the Great Lakes: An Environmental History, 1783–1933*. Madison, WI: University of Wisconsin Press.

Brantz, D. 2010. 'Introduction'. In D. Brantz (ed.) *Beastly Natures – Animals, Humans and the Study of History*. Charlottesville & London: University of Virginia Press. pp. 1–13.

Creager, A.N.H. and W.C. Jordan. 2002. 'Introduction'. In A.N.H. Creager and W.C. Jordan (eds). *The Animal/Human Boundary – Historical Perspectives*. Rochester, NY: University of Rochester Press. pp . ix–xviii.

Fagan, B. 2018. *Fishing – How the Sea Fed Civilization*. New Haven and London: Yale.

Forstén, A. 1902. *Kansan emännän keittokirja – sovitettu käytettäväksi myöskin yksinkertaisissa ruoanlaittokursseissa kansan naisille*. Helsinki: Helsingin sentraalikirjapaino.

Freedman, P. 2007. 'Introduction – a new history of cuisine'. In P. Freedman (ed.) *Food – The History of Taste*. Berkeley and Los Angeles, CA: University of California Press. pp. 7–33.

Friberg, A. 1900 [1890]. *Kansan keittokirja*. 2nd edition. Porvoo: WSOY.

Gottberg, G. 1913. *Kalastustutkimuksia Skandinaaviassa, Saksassa ja Itämeren maakunnissa*. Helsinki: Keisarillisen senaatin kirjapaino.

Greenberg, P. 2010. *Four Fish – the Future of the Last Wild Food*. New York: Penguin Press.

'H.A.' 1942. 'Kalan kuivaaminen kotioloissa'. *Emäntä-lehti* 5 (1942): 97–99.

Haapala, A. 1926. *Silakan suolauksesta*. Helsinki: Suomen kalastusyhdistys.

Hakamies, P. 2004. 'Ruisleipä ja muut ruokasymbolit'. In M. Knuuttila, J. Pöysä and T. Saarinen (eds). *Suulla ja kielellä – tulkintoja ruoasta*. Helsinki: Suomalaisen kirjallisuuden seura. pp. 79–91.

Häkkinen, K. 2002. 'Eläin suomen kielessä'. In H. Ilomäki and O. Lauhankangas (eds). *Eläin ihmisen mielenmaisemassa*. Helsinki: Suomalaisen kirjallisuuden seura. pp. 47–55.

Hannikainen, M.O. 2022. 'Roskaa vai ruokaa? Keittokirjojen kalat 1900-luvulla'. In J. Mikkonen, S. Lehtinen, K. Kortekallio and N-H. Korpelainen (eds). *Ympäristömuutos ja estetiikka*. pp. 196–235. https://doi.org/10.31885/9789526996103

Hellevaara, E. 1927. 'Kalastuselinkeinon kohottamisesta'. *Yhteiskuntataloudellinen aikakauskirja* 3–4.

Hirschfelder. G. 2001. *Europäische Esskultur – Eine Geschichte der Ernährung von der Stenzeit bis heute*. Frankfurt and New York: Campus Verlag.

Hoffmann, R.C. 2002. 'Carps, cods, connections – new fisheries in the medieval European economy and environment'. In M.J. Henniger-Voss (ed.) *Animals in Human History – The Mirror of Nature and Culture*. Rochester, NY: University of Rochester Press. pp. 3–55.

Ilvesviita, P. 1995. *Paaluraudoista kotkansuojeluun – suomalainen metsästyspolitiikka 1865–1993*. Rovaniemi: Lapin yliopisto.

Järnefelt, H. 1923. 'Pari sanaa kalastusoloistamme'. *Turun Sanomat.* 29 June.

Järvikalastuskomitea 1958. *Järvikalastuskomitean oSámietintö vuodelta 1953 ja sen täydennysosa vuodelta 1958: (oSámietinnöt I ja II).* Helsinki.

Järvi, T.H. 1932. *Suomen merikalastus ja jokipyynti.* Porvoo and Helsinki: WSOY.

Järvi, T.H. 1941. *Suomen Kalastusyhdistys 1891–1941.* Helsinki: Suomen kalastusyhdistys

Johansson, L. 1987. *Herkkuja muikusta mustekalaan.* Porvoo: WSOY.

Jutikkala, A. 2003. 'Katovuodet'. In V. Rasila, E. Jutikkala and A. Mäkelä-Alitalo (eds). *Suomen maatalouden historia I – perinteisen maatalouden aika esihistoriasta 1870-luvulle.* Helsinki: Suomalaisen kirjallisuuden seura. pp. 506–11.

Kalastuskomitea 1911. *Kalastuskomitealta.* Komiteamietintö 1910:7. Helsinki.

Kalataloudellinen toimikunta. 1951. *Kalataloudellisen toimikunnan mietintö kevätsilakan kalastuksen ja kaupan järjestämisestä.* Helsinki.

Kalatalouskomitea 1965. *Kalatalouskomitean mietintö.* Helsinki.

Kalateollisuustoimikunta 1985. *Kalateollisuustoimikunnan mietintö.* Helsinki.

Kalan markkinointitoimikunta 1974. *Kalan markkinointitoimikunnan mietintö.* Helsinki.

Kallioniemi, J. 2013. *123 sotavuosien ruokaohjetta.* Jyväskylä: Gummerus.

Kaski, L. 2019. *Myyttiset eläimet – tarua ja totta eläinten mahdista.* Helsinki: Suomalaisen kirjallisuuden seura.

Knuuttila, M. 2006. *Kansanomaisen keittämisen taito.* Helsinki: Suomen muinaismuistoyhdistys.

Knuuttila, M. 2013. 'Kun äiti jääkaapin osti'. In K-M. Hytönen and K. Rantanen (eds). *Onnen aika – valoja ja varjoja 1950-luvulla.* Jyväskylä: Atena Kustannus,pp. 173–81.

Koli, L. (ed.) 1984. *Suomen elämet 3.* Porvoo: Weilin+Göös.

Koli, L. 1990. *Suomen kalat.* Porvoo: WSOY.

Koli, L. 2001. *Otavan kalakirja.* 2nd edition. Helsinki: Otava.

Kolmonen, J. and V. Vanamo. 1973. *Maukasta kalasta.* Kuopio: Neuvontakokit.

Koskimies, H. and E. Somersalo. 1932. *Keittotaito – koteja ja kouluja varten.* Porvoo: WSOY

Kotimaisen kalan toimikunta 1990. *Kotimaisen kalan toimikunnan mietintö.* Helsinki: Maa- ja metsätalousministeriö.

Kunnas, H.H. 1914. *Äideille ja tyttärille – 222 valittua kasvi-, kala-, liha-, jälkiruoka- ja leivosohjetta y.m.* Hämeenlinna: Karisto.

Kylli, R. 2021. *Suomen ruokahistoria — suolalihasta sushiin.* Helsinki: Gaudeamus.

Laitinen, A. (ed.) 2014. *Suomessa kasvanut kala.* Helsinki: Suomen Kalankasvattajaliitto.

Lappalainen, A. 1995. 'Kalastuskulttuuri muuttuvassa yhteiskunnassa'. In L. Hyytinen and H. Kupiainen (eds). *Kalaveteen piirretty viiva – kalastus ja kalastaja yhteiskunnallisten muutosten pyörteissä.* Mikkeli: Helsingin yliopisto, maaseudun tutkimus- ja koulutuskeskus, pp. 53–87.

Lappi, L. (ed.) 1996. *Varituisia ja hapankaalia – vanhoja ruokatapoja ja tottumuksia Keurulla.* Keuruu: Keuruun maa- ja kotitalousnaiset.

Lehtinen E. and T. Salme. 1962. *Keittotaito – koteja ja kouluja varten.* 24th revised edition. Porvoo: WSOY

Levander, K.M. 1924. 'Merikaloista, joista valmistetaan kuivattua kalaa.' *Suomen Kalastuslehti* 6: 161–66.

Liimatainen, A. 2003 [1995]. *Parasta kotiruokaa.* 7th edition. Porvoo: WSOY

Lindstedt, A. 1997. *Pirkan parhaat kalaruoat.* 2nd edition. Helsinki: Kauppiaiden kustannus Oy.

Mela, A. E. J. and K. E. Kivirikko. 1909. *Suomen luurankoiset.* 2nd revised edition . Porvoo: WSOY

Metsästys ja kalastus. 1912. 'Milloin kalat ovat parhaita syötäviksi'. (9): 32.

Mink, N. 2009. 'Forum – it begins in the belly'. *Environmental History* **14** (2): 312–322.

Muuttuvien vesistöjen kalatalouden hoitotoimikunta. 1967. *Muuttuvien vesistöjen kalatalouden hoitotoimikunnan mietintö.* Helsinki.

Nenonen, M. 1938. *Silakka ja silli ruokataloudessa.* Helsinki: Maatalousnaiset.

Nissinen, A. and E. Somersalo. 1943 [1942]. *Kortiton ruoka ja miten käytän ruoka-annokseni.* 2nd edition. Porvoo: WSOY.

Nordlund, M. 1978. *Parhaat kalat ja kalaruoat.* Helsinki: Kauppiaitten kustannus Oy.

Nordqvist, O. 1891. *Ehdotuksia toimenpiteisiin kalastuksen kohottamiseksi.* (trans. by K.M. Levander). Helsinki.

Nordqvist, O. 1902. *Kalastustaloudellinen käsikirja* (trans. by K.V. Puuska). Kalastajain ja metsästäjäin kirjasto 1. Helsinki.

Olsonen, K. 1940. 'Mikä on paras kala?' *Kotiliesi* 10: 260–61

Olsoni, K. 1934. *Kalaruokia.* Helsinki: Suomen kalastusyhdistys.

Partanen, K. 2017. *Vedestä ruokapöytään – suomalainen elinkeinokalatalous.* Helsinki: Pro Kala.

Peltoniemi, K. 1984. *Taistelu kirjolohesta – muistelmia uuden elinkeinon, kalanviljelyn, alkutaipaleelta Suomessa.* Helsinki: Suomen lohenkasvattajain liitto.

Pohja-Mykrä, M. and S. Mykrä. 2007. 'Luonnonvaraiset eläimet sodassa ja sodan kohteina'. In S. Laakkonen and T. Vuorisalo (eds). *Sodan ekologia – sodankäynnin ympäristöhistoriaa.* Helsinki: Suomalaisen kirjallisuuden seura. pp. 143–92.

Pukkila, H. 2008. *Sattumasoppaa – pulavuosien parhaat palat.* Helsinki: Tammi.

Räsänen, T. 2021. 'Tyhjenevä maa – suhde luonnonvaraisiin eläimiin'. In E. Ruuskanen, P. Schönach and K. Väyrynen (eds). *Suomen ympäristöhistoria 1700-luvulta nykyaikaan.* Tampere: Vastapaino, pp. 267–91.

Reinilä, E. S. Calonius and V. Krank. 1908. *Kotiruoka – keittokirja kotia ja koulua varten.* Helsinki: Otava.

Reinilä-Hellman, E. 1925. *Silakkaruokia.* Helsinki: Suomen kalastusyhdistys.

Reinilä-Hellman, E. S. Calonius, V. Krank-Heikinheimo and R. Tennberg. 1938. *Kotiruoka – keittokirja kotia ja koulua varten. 14. uusittu ja lisätty painos.* Helsinki: Otava.

Rytkönen, A. 1929. 'Piirteitä pielaveteläisestä ja maaninkalaisesta kansanomaisesta kalojen käytöstä'. *Emäntä-lehti* 1: 16–21.

Sillanpää, M. 1999. *Happamasta makeaan – suomalaisen ruoka- ja tapakulttuurin kehitys.* Vantaa: Hyvää Suomesta.

Sisävesikalastustoimikunta 1976. Sisävesikalastustoimikunnan mietintö. Helsinki.

Sonck-Rautio, K. 2017. 'The Baltic herring as agents in the socio-ecological System in Rymättylä fisheries'. In T. Räsänen and T. Syrjämaa (eds). *Shared Lives of Humans and Animals – Animal Agency in the Global North.* London & New York: Routledge, pp. 119–131.

Suomalainen, R. and P. Muurinen. 1991. *Kotoisat kana- ja kalaruoat.* Porvoo and Helsinki: WSOY.

Suomen FAO-toimikunta. 1952. *Selostus v. 1951 tapahtuneesta kehityksestä maatalouden, metsätalouden, kalastuksen, ravitsemuksen ja kotitalouden aloilla.* Helsinki.

Suomen kalastusyhdistys. 1972. *Uutta ja hyvää kalasta – kalaruokakilpailun parhaita ohjeita.* Helsinki: Suomen kalastusyhdistys.

Svanberg, I. and A. Locke. 2020. 'Ethnoichthyology of freshwater fish in Europe: a review of vanishing traditional fisheries and their cultural significance in changing landscapes from the later medieval period with a focus on northern Europe'. *Journal of Ethnobiology and Ethnomedicine* **16** (68) https://doi.org/10.1186/s13002-020-00410-3

Talve, I. 1961. *Kansanomaisen ruokatalouden alalta.* Helsinki: Suomalaisen kirjallisuuden seura.

Tennberg, R. 1936. 'Ruodotonta kalaa'. *Kotiliesi* 11: 460.

Tennberg, R. and R. Rautiainen. 1960. *Kotiruoka: keittokirja kotia ja koulua varten.* 23rd edition. Helsinki: Otava

Toussant-Samat, M. 2009 [1987]. *A History of Food.* New expanded edition, trans. by Anthea Bell. Chichester: Wiley Blackwell.

Valle, K.J. 1934. *Suomen kalat.* Helsinki: Otava

Valtion kotitaloustoimikunta 1918. *Kalojen säilöönpaneminen kotitaloudessa.* Valtion kotitaloustoimikunnan tiedoksiantoja 38. Helsinki.

Vetikko, J., K. Nyberg and V. Rinne. 2009. *Kesämökin kalaherkut – ahvennapsikkaista rosvolahnaan.* Helsinki: WSOY.

Vuorisalo, T. and M. Oksanen. 2021. '"Mikä on toiselle hyödyksi, voi usein olla toiselle vahingoksi" – pohdintoja eläinluokittelusta'. In T. Räsänen and N. Schuurman (eds). *Kanssakulkijat – monilajisten kohtaamisten jäljillä.* Helsinki: Suomalaisen kirjallisuuden seura. pp. 23–49.

Walli, J. 1933. 'Kalojen ilmakuivatuksesta'. *Suomen Kalastuslehti* 3: 43–45.

Walton, J.K. 2000 [1992]. *Fish and Chips and the British Working Class 1870–1940.* Leicester: Leicester University Press.

Wartiainen, K. and K. Tolvanen. 1956. *Kotikokki keittää (kodin neuvokki 1).* Helsinki: Yhtyneet Kuvalehdet.

Yrjölä, S., H. Lehtonen and K. Nyberg. 2016. *Suomen kalat.* 2nd edition. Helsinki: Nemo.

Appendix

Table 2. The catch of professional fishers from Baltic Sea in 1972, measured by weight and value.

	Species	Amount (1,000 kg)	Share (%)		Species	Value (1,000 FM)	Share (%)
1	Baltic herring	53,758	85.1	1	Baltic herring	17,758	41.3
2	Whitefish	1,529	2.4	2	Atlantic salmon[1]	8,010	18.6
3	Perch	1,099	1.8	3	Whitefish	4,996	11.6
4	European sprat	972	1.5	4	Pike	2,588	6.0
5	Vendace	935	1.5	5	Bream	1,524	3.5
6	Pike	921	1.5	6	Herring	1,500	3.5
7	European smelt	832	1.3	7	Pike-perch	1,470	3.4
8	Bream	765	1.2	8	Burbot	1,466	3.4
9	Herring	570	0.9	9	Perch	1,454	3.4
10	Atlantic salmon[1]	456	0.7	10	Vendace	1,143	2.7
11	Miscellaneous	423	0.7	11	European sprat	553	1.3
12	Burbot	406	0.6	12	European smelt	220	0.5
13	Pike-perch	377	0.6	13	Miscellaneous	219	0.5
14	Ide	134	0.2	14	Ide	135	0.3
15	Cod	8	>0.1	15	Cod	12	>0.1
	Total	**63,185**	**100**		**Total**	**43,048**	**100**

[1] Includes Sea trout, Lake trout, Lake Salmon and other species of *Salmonidae* not otherwise mentioned.

Table 3. The catch of professional fishers from freshwater in 1972, measured by weight and value.

	Species	Amount (1,000 kg)	Share (%)		Species	Value (1,000 FM)	Share (%)
1	Vendace	2,777	75.6	1	Vendace	4,369	63.7
2	Whitefish	176	4.8	2	Whitefish	604	8.8
3	Pike	139	3.8	3	Pike	467	6.9
4	Miscellaneous	130	3.5	4	Atlantic salmon[1]	393	5.7
5	Perch	121	3.3	5	Burbot	254	3.7
6	European smelt	84	2.3	6	Pike-perch	219	3.2
7	Burbot	82	2.2	7	Perch	217	3.2
8	Bream	77	2.1	8	Bream	179	2.6
9	Pike-perch	49	1.3	9	Miscellaneous	85	1.3
10	Atlantic salmon[1]	26	0.7	10	Ide	39	0.6
11	Ide	13	0.4	11	European smelt	23	0.3
	Total	**3,674**	**100**		**Total**	**6,858**	**100**

[1] Includes Sea trout, Lake trout, Lake Salmon and other species of *Salmonidae* not otherwise mentioned. Source: Kalan markkinointitoimikunta 1974, p. 59.

SECTION 2

CONTESTED AND COLONISED SPACES

Chapter 6

CULTURAL NATURE IN MID-LAPPISH REINDEER HERDING COMMUNITIES

Maria Lähteenmäki, Oona Ilmolahti, Outi Manninen and Sari Stark

Finnish Lapland has often been seen as a pure, clean and attractive tourism area, one of the most popular European skiing and winter holiday centres, where the rich people from the south have luxurious wooden villas. On the other hand, northernmost Arctic Lapland has been seen as a romantic home region of the Mountain Sámi herders, where the indigenous people live a traditional life. These two stereotypical images have overshadowed one part of Lapland, which has faced the greatest environmental changes during the last hundred years. That area is the forest zone located in the middle of Lapland between the Arctic Circle and Sápmi, the Sámi cultural home area (Map 1). Alongside environmental changes, the region has faced an ethnic transition during the eighteenth and nineteenth centuries in which the Forest Sámi people have become one with the Finnish settler population. In mainstream ethnic and sociocultural studies related to Finnish Lapland, the region has been seen as an uninteresting borderland. Nevertheless, the multidimensional historical changes during the last three hundred years have formed the region into a peculiar cultural area,[1] which we call here the *Mid-Lappish hybrid cultural region.*[2]

According to our research material, reindeer herders living in Mid-Lapland identify mostly as carriers of historical Forest Sámi traditions, especially when they talk about nature and their close to nature way of life. Reindeer

1. Lähteenmäki 2006a, pp. 60–79.
2. 'Lappish' means all the people living in the province of Lapland, not only the Sámi. The concept was accepted in the 1950s.

doi: 10.3197/63824846758018.ch06

herding itself is locally seen as a manifestation of the unbroken chain to the Forest Sámi ancestors.[3]

> Our culture is based on inseparable connection with nature. We live out of nature, we respect nature, and we are part of nature. It is difficult to separate culture and way of life, they are intimately connected to each other. A lot is based on the old Forest Lapp culture. Reindeer herding itself is a culture of its own with its working methods and traditions. A lot of our cultural features have also survived until today.[4]

Research settings

Our research task is to present and analyse the features in the local human-nature and human-reindeer relations across the historical timespan of a century and in the context of cultural nature in the historical Forest Sámi area in Finnish Mid-Lapland. By the term *cultural nature* we refer to the different meanings and attributes groups and individuals give and have given to their surrounding natural environment with its fauna, flora and waterways. The question is viewed through environmental changes and the meanings connected to reindeer roundups (corrals) and roundup places as an example of human-nature interaction. The reindeer roundups have historically been important social meeting places among subarctic communities, and roundup events have been traditionally the highlight of the reindeer year. The events have been organised several times a year, usually during the early spring for sorting and counting reindeer calves, and larger events during the autumn-winter period for entire herd. An aspect related to the intergenerational nature of herding is the fact that the reindeer herding is based on private ownership, and the reindeer pass on in families through generations.

Nature is always a cultural construction that takes shape in people's minds, and humans process their place in the world in relation to the surrounding natural environment.[5] The so-called wilderness is part of the human realm as well. The terms natural or natural landscape are used of things modified by humans,[6] and reindeer herding with its corrals can be seen as one construction of the natural or ancient that is also cultural and relatively recent. Ideas concerning nature are inherited from previous generations, and they affect the

3. Cultural Nature Survey 2021: Respondent 1 (personal details not given).
4. Cultural Nature Survey 2021: Respondent 13, female, born 1987.
5. Richardson 2001, pp. 80–81.
6. Williams 1980, p. 79.

personal, embodied relationship with the natural environment.[7] Environmental historian Carolyn Merchant uses the concept of consciousness – that is, the ways in which a group collectively understands their surrounding nature. The community's collective consciousness is manifested in images of nature.[8]

Our research topic is framed by *environmental history*, and environmental humanities. The aim of this research field is to historicise the changing relationship and interaction between humans and nature, one which never had an ideal primordial state, as Melanie Arndt has argued. Nature and history are linked in a complex way, and environmental history is always also history of power.[9] This is true also in our study. Environment is 'that which surrounds us' (orig. French *environ*); in an extended sense it signifies the circumstances and conditions that make up everyday life.[10] The focus of environmental history research is the impact of human interaction with nature, both the intended and especially the unintended long-term consequences. In addition to the natural environment (the natural materials and living things researched by biologists and environmentalists), the sociocultural, the economic and the political (societal) environments – which we call here humanised environment – are everything around us. These kinds of environments are the conditions in which people live, work or spend time, and influence how they feel, behave or work. People's beliefs and actions depend on environments: by studying historical environments, one studies the interaction between humans and environments in the past and in recent history.[11]

Reindeer husbandry has been an important part of the economic structure of communities in Lapland at least since the eighteenth century when the number of reindeer started to increase in Northern Fennoscandia. Before that, wild reindeer were hunted for their skins, antlers and meat. They were also used as beasts of burden and as bait animals when hunting wild deer.[12] Human-reindeer coexistence has deep historical roots in the process where reindeer as game developed to the semi-domesticated animals they are today. The reindeer husbandry organisation faced a dramatic turn in Finnish Lapland in 1898 when the state government, the Finnish Senate, took reindeer husbandry under its

7. Lähteenmäki et al. 2019; Laurén 2006; Smith 2006.

8. Merchant 1987, p. 272, passim.

9. Melanie Arndt, Environmental History 2016: https://docupedia.de/zg/Arndt_environmental_history_v3_en_2016 (accessed 16 June 2021).

10. E.g. Hughes 2008.

11. Hughes 2006; McNeill 2003; Isenberg 2014, passim.

12. Kortesalmi 2008, passim.

control, and divided the northernmost region of Finland into reindeer herding districts – that is, cooperatives.[13] The Finnish reindeer herding region in 2020 contained 54 cooperatives of which 31 were located beyond the Arctic Circle.[14]

Our empirical focus lies in two *reindeer herding cooperatives* (Finn. *paliskunta*), Sattasniemi and Oraniemi, geographically located in the middle of Finnish Lapland – mainly in Sodankylä, and partly in Savukoski and Pelkosenniemi municipalities (Map 1) – and the reindeer roundup processes in these cooperatives.[15] Our key source data consists of archival material, such as the minutes and reports of the Reindeer Herders Association and Sattasniemi cooperative located in the Finnish National Archives. We have also utilised regional, local and occupational newspapers and magazines from the 1920s to the 2010s. In order to reach the voices of the contemporary herder communities we conducted a *Cultural Nature Survey* during the period from 22 February to 30 March 2021. We had fourteen respondents, of whom sixty per cent were male and forty per cent female. They were born between 1950 and 1987. Fifty-six per cent of the respondents work full time in reindeer herding and 36 per cent part time. Twelve respondents were from Sattasniemi and two from the Oraniemi cooperative. In this historical study, we have analysed the sources by using qualitative methods, close reading[16] and survey analysis.

13. The reindeer herding law was drafted in Finland first in 1932, remodified in 1948 and 1968. The latest law was enacted in 1990.

14. Paliskunnat 2020. Paliskuntien yhdistys: https://paliskunnat.fi/py/paliskunnat/paliskuntien-tiedot/ (accessed 15 Nov. 2022).

15. The article is a part of the HISTECO project funded by the Academy of Finland, University of Lapland and University of Eastern Finland.

16. Close reading as a method is the careful, sustained interpretation of historical texts which emphasises the single and the particular over the general; working with historical texts requires a specific kind of conversation between reader and text. This conversation begins with the previously discussed questions, which ask the reader to read 'around' the historical text, noting time, place, authorship and other information about the text's origin and historical context. It continues with a close examination and conversation with the text itself. See, for instance, Close Reading and Annotating Historical Sources: https://lessonresearch.net/content-resource/close-reading/ (accessed 16 June 2021).

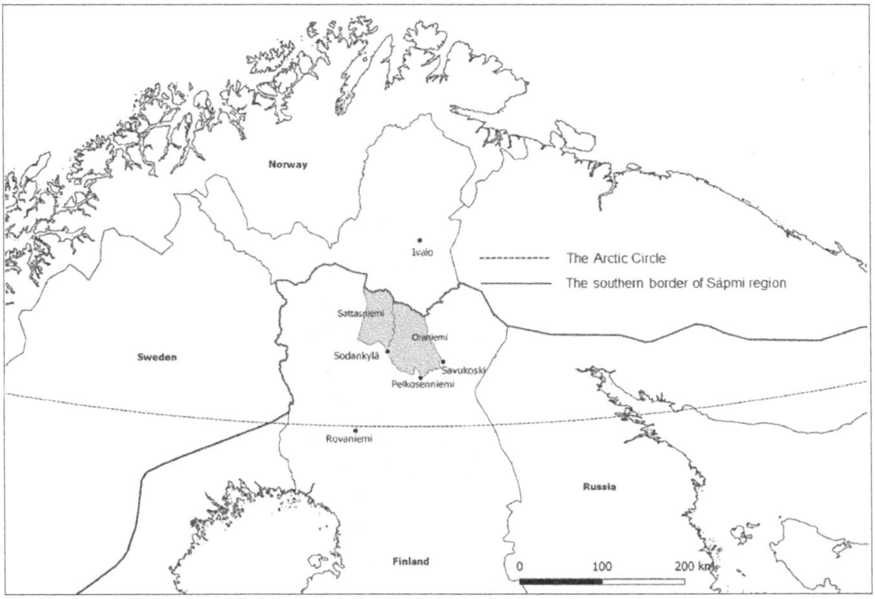

Map 1. The research area in the Middle of Finnish Lapland, beyond the Arctic Circle.

Sources: The map contains data from the Finnish Environmental Institute (TOKAT database), National Land Survey of Finland (Topographic database 2019), and Global Aviation Data Management (GADM database 2015). Drawing by Outi Manninen 2021.

In the middle can be seen the geographical and cultural research area. Our research area was originally a historical administrative region of the Large Sodankylä community, which existed until 1916, and formed a coherent cultural region. The geographical district is vast: the acreage of Sodankylä-Savukoski region alone is nearly 19,000 km^2, with a population density of 0.16–0.71 persons/km^2 (Table 1).[17]

17. In 2020 there were 8,300 inhabitants in Sodankylä, of whom 1.6% spoke Sámi; in Savukoski there were 1,010 inhabitants, of whom 0.4% spoke Sámi. The number of inhabitants in Pelkosenniemi was 930.

Table 1. The research region in numbers.

Cooperative	Municipality in Lapland	State-owned land %	Acreage km^2	Number of reindeer, max	Number of reindeer owners	Number of roundups, 2021	Nature protected areas (forests, swamps, peatlands, groves)	Mines	Artificial lakes
Sattas-niemi	Sodankylä	94	2432	5300	162	14	Pomokaira, Kaarestunturi Kaaresvuoma Tenniöaapa Ilmakkiapa	Pahtavaara	5
Oraniemi	Sodankylä 60%; Savukoski 25%; Pelkosen-niemi 15%	75	3893	6000	130	35	Viiankiaapa Koitelainen Kyläselkä Ellitsa Nivatunturi Lämsänaapa-Sakkala Luiro swamps Leviäaapa-Sammalaapa	Kevitsa, Sakatti (planned)	
Total		85	6325	11,300	292			3	5

In the seventeenth century the Forest Sámi, who differed from the other Sámi groups in their language and culture, populated the Mid-Lapland region. Mainly due to influence of the Finnish administration and mixed marriages, the old Forest Sámi language died out in the nineteenth century, and the herders of these cooperatives no longer wear the traditional Sámi costumes in their everyday work. Typically, old photographs from the turn of the nineteenth and twentieth centuries presenting life in Mid-Lappish villages introduce only the few Sámi people who were originally Mountain Sámi and moved to Mid-Lapland from Northern Norway in the late nineteenth century.[18]

Regarding the historical ethnic adaption process, the borders of the Sámi area in Lapland were officially opened to settlers in 1673, and the Mid-Lappish

18. Photographs in the Finnish Heritage Agency's Database: keywords 'saamelaiset Sodankylä': https://museovirasto.finna.fi/ (accessed 20 May 2021). On the hierarchy of Sámi people in Finnish Lapland, the lost Forest Sámi culture, and the daily life of the pioneer settlers in Sodankylä, see Lähteenmäki 2006b, pp. 191–212, 163–179.

region became mixed Finnish-Forest Sámi during the next two centuries. Even today, however, reindeer herding is the main source of living for many families in our research area. Due to the cultural change, these families, who called themselves as *descendants of the Forest Sámi* have not been accepted on the Electoral list of the Sámi Parliament in Finland as 'real' Sámi people, unless they have documentary proof of adequate Sámi ancestors. During the last ten years, the cultural place and space of these so-called 'non-status Sámi', 'people in between' and 'Finns who have Sámi roots' – has been the subject of hot debate in ethnic and identity discourses and the revitalisation process of the Forest Sámi culture in Mid-Lapland.[19] In spite of the cultural denial, the strong intergenerational link through the centuries still influences local living culture and reindeer husbandry processes in the reindeer cooperatives of this study.

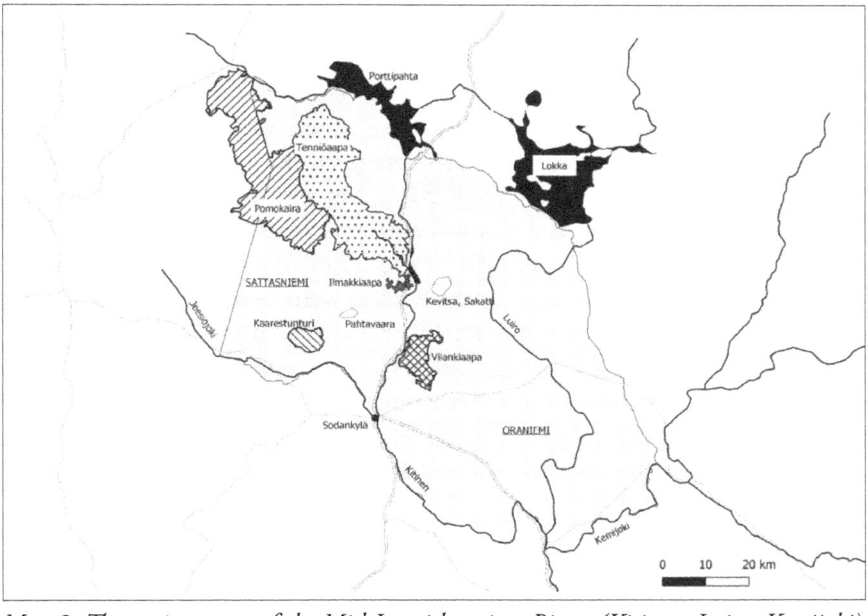

Map 2. The environment of the Mid-Lappish region: Rivers (Kitinen, Luiro, Kemijoki), artificial lakes (Porttipahta, Lokka), mires (Tenniöaapa, Ilmakkiaapa, Viiankiaapa), wilderness area (Pomokaira), mines (Pahtavaara, Kevitsa, and planned mine Sakatti), fell (Kaarestunturi) in the Sattasniemi and Oraniemi cooperatives.

Sources: The map contains data from the Finnish Environmental Institute database (TOKAT database and nature protected areas), National Land Survey of Finland (Topographic database 2019), and GADM (database 2015). Drawing Outi Manninen 2021.

19. E.g. Sarivaara and Uusiautti 2013.

The concept *Forest Sámi* originates from dominantly forested nature of Mid-Lapland. In addition to forests, the local nature consists of mires, such as Pomokaira, Tenniöaapa and Viiankiaapa, and rivers. Kemijoki is the longest river in Finland, and its tributaries, Kitinen, Luiro and Jeesiöjoki, all run through our research area. Settlements, small villages, are located alongside the waterways. In the north, the cooperatives border two artificial lakes, Lokka and Porttipahta. Lokka is the largest artificial lake in Europe (Map 2). Large-scale forestry since the end of the nineteenth century, and energy production, the mining industry, motorisation, tourism and even nature conservation after World War Two have all affected and altered the local natural environment.

In our data there are several features that indicate an identification with the *Forest Sámi way of life*. The first identification argument of our respondents is related to the local livelihood itself. From the late seventeenth century,[20] Forest Sámi hunter-gatherer culture gradually merged with Finnish peasant culture, and the area became a hybrid cultural region; reindeer herding maintained an important part in this subarctic environment among the Forest Sámi, but also among the newcomers, Finnish cattle-raising farmer-settlers. They adopted small-scale reindeer herding from the Sámi. Since then, this combination of livelihoods has characterised the mixed economic system of the Mid-Lapland villages. In 1912, some 45 per cent of the population in Sodankylä owned reindeer, and over sixty per cent of the herders were landowning peasants.[21]

Another feature that connected them to their pastoral past was 'timeless time perception'.[22] This is one reason for the importance of the roundup events. It has been said that it is more natural to live according to nature than to the clock.[23] The descendants of the Forest Sámi still see the natural environment as an organic part of their work and culture, as a young reindeer herder woman puts it:

> Reindeer tending work is based on the cycle of nature: calf markings, *etto* [gathering of animals], feeding the reindeer, transporting the reindeer, watch-

---

20. The settlers arrived in Sámi villages in Sodankylä-Savukoski after settlement legislation in 1675 and 1695, which opened the Lapland border to Finnish permanent settlers.

21. Heikel et al. 1914, pp. 190–91.

22. Female author, born in 1986: Instagram 11 Feb. 2020.

23. Interview with Iida Melamies. See: https://yle.fi/aihe/artikkeli/2021/05/07/iida-melamies-haluaa-vaikuttaa-poronhoidon-tulevaisuuteen-ja-nyt-nuori-nainen (accessed 27 May 2021).

The border fence between reindeer herding cooperatives in Mid-Lapland. Photo: Maria Lähteenmäki 2021.

> ing the reindeer, mending, circling, and building the corrals. All year around observing nature, exercising, hiking, fishing, picking berries and mushrooms.[24]

A third connection between today's inhabitants and the past Forest Sámi, is the working sites and homes: 'We are living in the old dwelling places of the ancestors'.[25] A respondent from Oraniemi emphasised that his ancestors had herded reindeer in his village for at least 250 years, and still in the 1980s

24. Cultural Nature Survey 2021: Respondent 13, female, born 1987.

25. See Metsälappalaiset. Mitä on metsälappalaisuus? https://www.metsalappalaiset.net/mita-on-metsalappalaisuus (accessed 15 May 2021); Antikainen et al. 2019, pp. 33, 43.

almost everyone owned reindeer.[26] A younger female herder from Sattasniemi described how she was born to her occupation; her family has worked with reindeer for hundreds of years, 'so it's almost like hereditary for us'.[27] Our survey illustrates the herders' mental orientation towards the traditional Forest Sámi hunter-gatherer lifestyle.[28] Self-sufficiency, an outdoor lifestyle, berry picking, hunting and fishing are repeatedly brought up in the answers as cornerstones of both leisure time and livelihood.[29]

The changing environment in Forest Lapland

When asked how local nature has changed, our survey respondents brought up visible transitions in the natural environment that are mostly due to the acts of 'outsiders', such as the national administration, especially the National Forest Service (Finn. *Metsähallitus*), private business companies or decision-makers in the 'South'. The biggest problems perceived for reindeer herding were forest industry and logging (86%), artificial lakes (86%), and mining (71%).[30] The three eldest respondents, born in the 1950s and 1960s, emphasised the rise in the number of forest roads,[31] and the decrease in untouched wilderness. The loss of peaceful, dense forests seems to be experienced as being just as painful as the disadvantages these industries have created for reindeer husbandry, which uses the same forests as pastures. According to the respondents, the scenery has been transformed from the 1960s onward because of mining, test drilling[32] and wind turbines, also called 'devil's whisks'.[33] A certain dualism can be seen

26. Cultural Nature Survey 2021: Respondent 10, male, born 1965.
27. YouTube. #vaihtovuosisodankylässä vlog 28 Feb. 2020: https://youtu.be/WHhW4Ok-hACA (accessed 25 May 2021).
28. Cultural Nature Survey 2021: Respondent 6, male, born 1970; Respondent 7, female, born 1986; Respondent 12, female, born 1982.
29. Cultural Nature Survey 2021. When asked what the respondents do in nature, fishing is mentioned in 12/14 answers, hunting in 9, berry picking in 8 and mushroom picking in 1 answer. Working with reindeer in the forest is understood as part of this self-sufficient lifestyle.
30. Cultural Nature Survey 2021.
31. Cultural Nature Survey 2021: 'Excessively dense forest road network'. Respondent 2, male, born 1953; 'Forest roads'. Respondent 4, male, born 1950; 'Roads built everywhere'. Respondent 5, male, born 1965.
32. Motorisation and technical utilities (71.4%) come only after the profound environmental changes. Cultural Nature Survey 2021.
33. Cultural Nature Survey 2021: Respondent 3, male, born 1953.

in the respondents' descriptions of local nature: the ideal scenery is beautiful and nostalgic with ancient spruces, old forests, open swamp landscapes and wilderness. Pomokaira nature reserve was seen to form a contrast to the rest of the cooperative's area: clearcut commercial and ravaged forests.[34]

> Wish there were clean waters and uncut forests. Freely running clear water is the source of life, of which man should never give up... On our lives ... the harnessing of the river and building the artificial lakes has had an enormous effect. The Kitinen was a beautiful natural river with big and small rapids and steam pools. ... Now what is left is a dug-up hole, which doesn't freeze even for the winter because of the regulated running... The wilderness of my childhood has changed a lot.[35]

The visible consequence of intensive forestry in Mid-Lapland has been the decrease of old trees with beard moss (Finn. *naava*), lichen growing in spruces, which is an important reserve nutrition for the reindeer in the winter.[36] Cutting the forest has brought other disadvantages, as Oraniemi cooperative reported in 1959: 'Winter pastures have diminished quite a bit due to the forest cutting carried out by the government. There is so much fallen wood on the ground that reindeer are not able to dig in the snow season.'[37] The consequences of forestry have been multidimensional: diminishing beard moss and lichen areas, moulding the terrain, fallen trees in the forests, hardening snow in open spaces, difficulties in moving, disturbances to the pasture rotation and scattering of the animals.[38]

In the 2020s, the herding processes of both Oraniemi and Sattasniemi cooperatives are still based on widespread use of natural pastures. The summer pastures, such as mires and twig and grass pastures, provide enough nutrition for the animals, but the winter lichen pastures have diminished, and therefore supplemental feeding has become a necessity. Supplementary feeding started

---

34. Cultural Nature Survey 2021: Respondent 3, male, born 1953, and Respondent 8, female, born 1975: 'Everything is excavated, modified and cut.' Respondent 7, female, born 1986.

35. Female author, born in 1986, on Instagram 10 Aug. 2020. The citations have been translated by the authors.

36. See Paliskuntain yhdistys. Challenges in Reindeer herding: https://paliskunnat.fi/poro/poronhoito/poronhoidon-haasteet/ (accessed 26 March 2021); Kumpula, Kurkilahti, Helle and Colpaert 2014.

37. Annual report of Oraniemi cooperative 1958–1959. Reindeer Breeding Association's Archives, Eb:3. NA-Oulu.

38. Kumpula and Nieminen 1987, p. 27.

in the cooperatives in the 1960s, at first on a small scale.[39] Twenty years later, fifteen per cent of reindeer in the cooperatives were held in yards.[40]

The most profound changes in the natural environment occurred after the Second World War, although expanded industrial forestry had altered the landscape from the late nineteenth century, when stock rafts began to be floated along the Kemijoki River and its tributaries to the paper and pulp mills in Kemi on the coast of the Gulf of Bothnia. Later, the construction of Lokka artificial lake (opened in 1969) for Finnish energy production had remarkable effects on the work of the reindeer herding cooperatives in Mid-Lapland (Map 2). As early as 1958, reindeer herders claimed that the first forest cutting to make way for the artificial lake had ruined most of the reindeer pastures, and its destructive influences on their livelihood were continuously referenced until the early 1970s. The lack of pasture had led to reindeer wandering to the neighbouring cooperatives, and they could no longer be herded in stocks (herds, Finn. *tokka*) in winter. The situation also affected the nutrition of reindeer and led to their deaths.[41] One cooperative in the area lost some forty per cent of its summer pastures due to the Lokka lake, and century-old traditions of dock herding and leash calving came to a sudden end.[42] Nature protectors also highlighted the loss of unique nature values, with irreplaceable damage being done to the local natural environment.[43]

The Porttipahta artificial lake (Map 2) opened in 1970. In May 1964 the Sattasniemi cooperative raised the fact that the Kemijoki Company had already applied for a building permit to construct a new artificial lake. However, this fact did not generate much discussion, although a working group was founded to deal with the lake issue and demanded compensation for the damage.[44] It was stated in 1968 that there was a great need for a fence between Sattasniemi and Kuivasalmi in Kittilä cooperatives, because the Porttipahta cutting area had made the reindeer wander further from the cooperative's border, and run-

---

39. Sakatti Mining Company. Porotalousselvitys [Reindeer herding economy report] 25 Nov. 2020. The first references to supplementary feeding in Sattasniemi cooperative are found in 1966. See Sattasniemi cooperative meeting 30 Dec.1966 §6 and §8. Sattasniemi cooperative archives, Ca:6. NA-Oulu.

40. Kumpula and Nieminen 1987, p. 29.

41. Annual reports of Lappi cooperative 1957–71. Reindeer Breeding Association Archives, Eb:3–Eb:6. NA-Oulu; Lappi cooperative is situated in Sodankylä as well, to the north of Oraniemi cooperative.

42. Pyhäjärvi 2011, pp. 32–33.

43. Rauno Ruuhijärvi, 'Hukkuvaa Sompion Lappia'. *Suomen Luonto* 3/1959, p. 677.

44. Sattasniemi cooperative Spring meeting 4 May 1964, §14–15. NA-Oulu.

ning the water to the reservoir would make things even worse.[45] Sattasniemi cooperative had border fences and roundup corrals in the Porttipahta lake area, which it had to abandon,[46] and the regular running of water was predicted to cause reindeer deaths.[47] The lake has been said to ruin some of the best pastures in the area of the Sattasniemi cooperative.[48] The cooperatives were not satisfied with the one-off financial compensation from the national Hydropower Committee, and the legal process proceeded through all the courts.[49]

In addition to the artificial lakes and their effects on the local environment, the Kelukoski hydropower plant built beside the Kitinen River in 2001 changed the surroundings.[50] In the 2000s, wind farms have been actively debated. Four herder women from Sattasniemi accused the council of Sodankylä municipality of pursuing a 'fast buck' and sacrificing the oldest livelihood and the most committed residents in the area: 'The reindeer keep us here generation after generation and make the young people come back after their studies.'[51] The former cooperative's chair called the wind farm the environmental crime of the century.[52] Locals emphasised that the vibrant work of the cooperative was based on natural pasture rotation, which yet another wind farm would jeopardise.[53] The planned windmills were seen to change the sense of place.[54]

45. Sattasniemi cooperative board meeting 16 Dec. 1968 §9. Sattasniemi cooperative archives, Ca:6. NA-Oulu.

46. Sattasniemi cooperative general meeting 17 Sept. 1962 §21, and Sattasniemi cooperative meeting 19 May 1970 §12. Sattasniemi cooperative archives, Ca:6. NA-Oulu.

47. Sattasniemi cooperative board meeting 8 Dec. 1973 §8. NA-Oulu.

48. Annual report of Sattasniemi cooperative 1964–65. Reindeer Breeding Association Archives, Eb:5. NA-Oulu.

49. Pyhäjärvi 2011, pp. 32–33; Reindeer Herders' Association annual meeting 4–5 June 1974, §14. Finnish Reindeer Herders' Association archives, Da:3. NA-Oulu; Sattasniemi cooperative board meeting 19 Aug. 1968 §3 and Sattasniemi cooperative general meeting 19 Sept. 1969 §9. Sattasniemi cooperative archives, Ca:6. NA-Oulu.

50. Cultural Nature Survey 2021: Respondent 13, female, born 1987; The homepage of Sattanen village: https://sattanen.info/wordpress/?page_id=53 (accessed 27 May 2021).

51. Kirsi Mäkitalo, Jenni Kaaretkoski, Iida Melamies and Viola Ukkola. 'Pikavipillä pinteestä porollisten kustannuksella?' *Lapin Kansa* 25 Feb. 2020.

52. Auvo Autio, 'Paikallisia on kuunneltava'. *Lapin Kansa* 12 March 2020.

53. The letter was published in the local magazine Sompio 25 Feb. 2020.

54. Cultural Nature Survey 2021: Respondent 13, female, born 1987.

However, reindeer herding was even more disrupted in the Oraniemi cooperative, due to the Kevitsa mine and snowmobile routes. [55]

The respondents to our survey claimed that the wind turbines, artificial lakes and mines have also altered the pasturing behavior of the animals. The building of the Pahtavaara gold mine from the 1980s onwards substantially decreased the use of the nearby Sattasvaara roundup corral, when reindeer started to avoid the area.[56] The Kevitsa mine also influences work in Oraniemi cooperative: reindeer have avoided this area too and changed their routes. The planned Sakatti mine would potentially jeopardise the whole livelihood in Oraniemi cooperative, causing damage comparable to the artificial lakes built in the 1960s and 1970s by overrunning and destroying a significant part of the pasturing lands.[57]

The highly emotional talk about changes in the local environment tells us a lot about the local human-nature relationship. It has been argued that the most valuable assets of any traditional community are its lands and its culture.[58] This is true also in the Mid-Lappish communities. Emphasising the close interaction between nature and cultural codes is crucial, even if both nature and humans' way of life have changed. A young respondent to our survey missed the untouched forests, which have been lost, as she had heard about them from the older members of the community. The ideal of pure and free nature is placed in the past, when nature was not exploited so much, while nowadays traces of humans are too visible.[59] The relation between nature and humans has been, according to the locals, unbalanced.

Nature protectors?

There is also another side of the coin when debating the question of preservation and nature protection. The members of the reindeer herding communities have identified themselves as nature protectors, but at the same time want to

---

55. Sakatti Mining Company. Porotalousselvitys [Reindeer herding economy report] 25 Nov. 2020, pp. 73–74, 97, 137–138.
56. Anttonen Marja, 'Porotalous ja muu maankäyttö: käsittelyssä kaivokset', *Poromies* 3/2011, pp. 41–43.
57. Sakatti Mining Company. Porotalousselvitys [Reindeer herding economy report] 25 Nov. 2020, pp. 137–38.
58. See Pilgrim and Pretty 2010, p. 1.
59. Cultural Nature Survey 2021: Respondent 13, female, born 1987.

distinguish between themselves and the green 'daydreamers of the South', who have no practical knowledge of living with nature (as became the perception as early as 1933 in terms of animal welfare).[60] At the meeting of the Reindeer Herders' Association in June 1952, the attitude toward nature conservation was positive, because the areas were supposed to become sanctuaries for wild animals, such as reindeer. However, the herders emphasised that they must have the full right to stay in conservation areas, take firewood and other necessities from the forest, and be able to overnight in the state-owned wilderness cabins. In addition, they demanded that they should be allowed to eradicate predators from the area, and all the old usufructuary rights should stay as they were.[61] Local autonomy in the use of the state-owned forest has been an important long-term issue when discussing nature protection in the Lappish communities. Nature conservation areas have been welcomed at least in principle, because they are good summer pastures for reindeer.[62] The same distinction is still made today; a female herder in our survey considered herself a nature protector, but she did not support the same values as the 'greens in the cities'.[63]

However, attitudes can change. Forest cutting has brought income to the area, and the regulations concerning nature conservation might have been experienced as a threat, as we can see in the discussion related to the Pomokaira conservation area. In 1976 the board of Sattasniemi cooperative discussed the National Forest Service's plan to to clear cut the Pomokaira, and unanimously concluded that it could be done.[64] A couple of months later, the cooperative board decided to strongly object to the proposed protection of the entire Pomokaira wilderness, although protecting the swamps was supported.[65] Also, the general meeting of the cooperative 'strongly and unanimously' opposed the protection of the area.[66] The reasons for objecting were likely connected

60. 'Eläinsuojelua koskeva lakiesitys eduskunnalle', *Lapin Kansa* 12 Sept. 1933.

61. Reindeer Herders' Association annual meeting 5–7 June 1952 §8 and §29. Finnish Reindeer Herders' Association archives, Da:1. NA-Oulu.

62. See, for example, Reindeer Herders' Association annual meeting 5–7 June 1952 §8 and Reindeer Herders' Association general meeting 25–26 May 1972. Finnish Reindeer Herders' Association archives, Da:1– Da:3. NA-Oulu.

63. Cultural Nature Survey 2021: Respondent 7, female, born 1986.

64. Sattasniemi board meeting 28 June 1976 §4. Sattasniemi cooperative archives, Ca:6. NA-Oulu.

65. Sattasniemi board meeting 10 Aug. 1976 §3. Sattasniemi cooperative archives, Ca:6. NA-Oulu.

66. Sattasniemi cooperative Autumn meeting 28 Sept. 1976 §2. Sattasniemi cooperative archives, Ca:6. NA-Oulu.

to the question of carnivores, as we can conclude from an earlier Koilliskaira and Koitelaiskaira protection area plan, which was feared in the early 1970s would constitute a 'beast pound' right at the edge of the cooperative.[67] This encapsulates the complex relationship local herders have with nature protection. An anecdote from the Reindeer Herders' Association meeting in 1975 related to nature reserves and the predator issue sums up the dilemma: 'We reindeer men are nature protectors, but not by any means enthusiasts.'[68]

One aspect of nature conservation is the damage reindeer do in the forests, an issue that has been discussed for over a century. Herding has conflicted with both agriculture and forestry because freely wandering animals are nowadays also seen as a threat to the environment. Earlier the focus was on economic losses for the forest industry and farmers and, from the 1970s, also increasingly on issues related to nature conservation.[69] This has led to strict quotas on reindeer numbers in each cooperative. Already in the early twentieth century, the 'golden years' of the Finnish timber industry, reindeer herding was opposed by the government and the professional forest industry, and reindeer were alleged to ruin state forests, for example by digging lichen and rubbing their antlers, a presumption reindeer herders claimed to be false.[70] The impact of herding on climate change today is also under discussion.[71] Nature conservation and a pastoral lifestyle have been seen as an impossible match, but lately reindeer herders are being considered as potential producers of ecosystem services.[72]

---

67. Sattasniemi cooperative Spring meeting 12 May 1971 §21. Sattasniemi cooperative archives, Ca:6. NA-Oulu.

68. Reindeer Herders' Association annual meetings 4–5 June 1974 §8 and 4–5 June 1975 §6. Finnish Reindeer Herders' Association archives, Da:3. NA-Oulu.

69. Heikkinen 2012.

70. See for example Heikel et al. 1914, pp. 89–104; J.H. Herva, 'Porotalouden perus, miten on rakennettava'. *Poromies* 1/1931, p. 3; Reindeer Herders' Association annual meeting 5-7 June 1952 § 9, § 29 and 3-4 June 1959 § 12; Reindeer Herders' Association annual meeting 5–7 June 5–7, 1952 §26. Reindeer Herders' Association archives, Da:1, NA-Oulu.

71. Järvenpää Juha, Poro ja poronhoito talousmetsissä: Katsaus metsätalouden ja porotalouden yhteensovittamiseen Suomessa [Report: Reindeer and reindeer husbandry in household forests] 2018, pp. 33–34.

72. Heikkinen et al. 2012, passim; see also Koster et al. 2013 on vegetation damage in Oraniemi cooperative.

Roundups as intergenerational human–nature interaction

Everything in reindeer husbandry in Mid-Lapland focuses on roundup corrals, where animals are gathered from the forests. Roundup sites are workplaces, social meeting sites and a cultural layer in the forest, where the work of the previous generations intertwines with the nature experiences of present-day herders:

> To the big roundups in the twentieth century a crowd of thousands of reindeer was brought from many directions: finally, the leads were full of antlered animals: the clicking of hooves, the rattling of horns, and grunting filled the air. And when the reindeer hypnotically circled in the corral round, round, round, running, 10,000 reindeer at once, the noises were merged into one whoosh and mixed in the fuming air.[73]

The oldest known corrals in our research area date back to the nineteenth century, but the same places might have been used in earlier centuries too. Many of the old roundup sites are still in use, and five traditional roundup corrals have been musealised, or partly protected as locally prominent cultural heritage. One of the museum corrals is Saarivaara roundup corral located in Oraniemi cooperative near the Luirojoki River. Massive roundups were still held in Saarivaara in the early twentieth century. The corral was musealised by the Finnish Heritage Agency (Finn. *Museovirasto*) in 1990–1992.

The preserved Saarivaara roundup in Oraniemi cooperative. Photos: Maria Lähteenmäki 2021.

In our survey, almost all old roundup corrals were found worthy of protection. Preservation of the corrals was justified for historical, cultural,

73. Strand 1994, p. 20.

ecological and practical reasons. Roundup places are considered important because the locations have been selected based on knowledge of the weather, nature, terrain and pasturage circle of reindeer; and due to the usefulness of the sites over the centuries. Information from earlier generations, called traditional ecological knowledge,[74] has guided herders to build corrals in the best spots in relation to natural pasture rotation.[75] Because of this, the wooden constructions are an important part of local cultural nature: the roundup places are unique landmarks of human-nature interaction. In the old roundup corrals a special kind of vegetation has emerged, such as birch groves and specific ground vegetation, so it is not just the man-made corrals that are visible signs of human action. The corrals are also historical landmarks of reindeer herders' knowledge of their natural environment and animal behaviour.[76]

In Mid-Lapland, many old corrals are still usable, and regularly repaired. The two respondents from Oraniemi cooperative mentioned four roundup sites that are important in terms of cultural nature.[77] Their main corral is Kyläselkä, located near the archaeological ruins of the last Sámi winter village (*siida*).[78] The other old corral, Routusvaara, is still in active use, and dates back to the nineteenth century, as can be observed from the tall old trees inside the area; outside the corral the vegetation is young coniferous pine forest.[79] Next to the Routusvaara corral are remains of the old roundup churn (Finn. *kirnu*).[80] Herders in Sattasniemi cooperative still use four historical roundup sites. The two oldest corrals, Kautoselkä and Salmuri, are mentioned in the survey as the most important sites in terms of cultural nature. The Salmuri corral[81] is located

74. Native peoples possess an extensive and deep understanding of their local ecosystem, in ethnoecology and more specifically traditional ecological knowledge. See Gadgil et al. 1993, pp. 151–56.

75. Cultural Nature Survey 2021: Respondent 5, male, born 1965; Respondent 6, male, born 1970; Respondent 7, female, born 1986; Respondent 8, female, born 1975; Respondent 12, female, born 1982; Respondent 13, female, born 1987.

76. Stark et al. 2022, passim.

77. Nuolikuru, Palomaa, Routusvaara and Tulinenjärvi corrals. Cultural Nature Survey 2021: Respondent 9, male, born 1970; Respondent 10, male, born 1965.

78. Piltz Martti, Selvitys Seitaniemen tien museoarvosta 2009: https://www.mobilia.fi/sites/default/files/seitajarvi_selvitys.pdf (accessed 7 March 2021).

79. Reindeer round-up in Orajärvi Routusvaara, 11 Oct. 2016. Mikko Maijala talks about the round-up in Routusvaara.

80. See metsa.fi: PAVE database, National Forest Service. PAVE is a geographical information system on structures, trails and archaeological sites.

81. Built in 1901 as a calf-marking corral. Sattasniemi cooperative general meeting 1 Oct. 1901 §18, NA-Oulu.

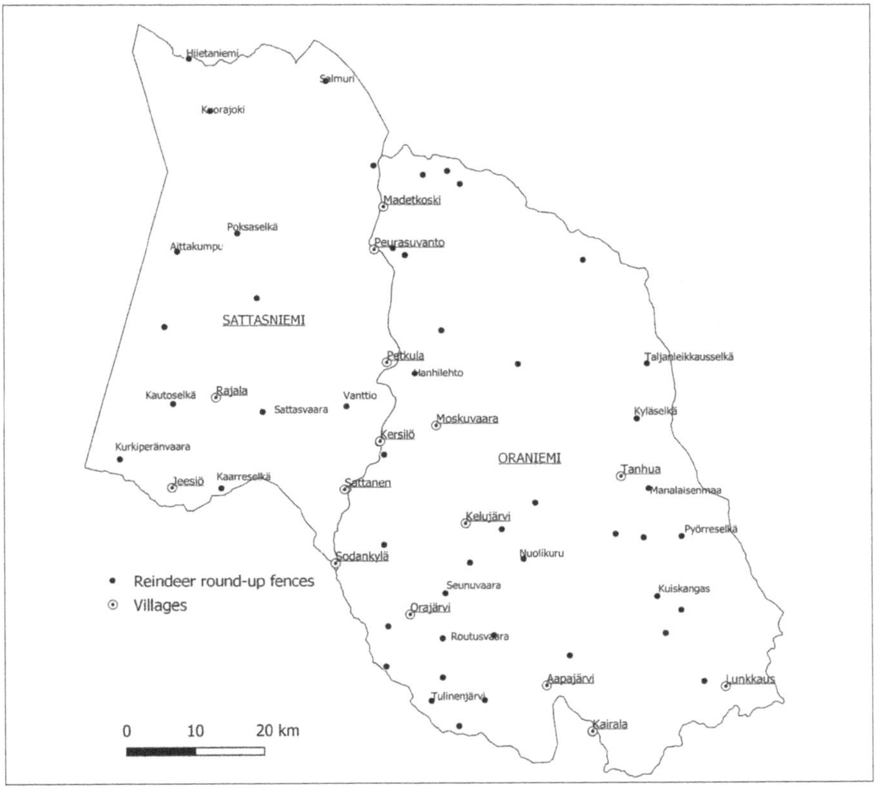

Map 3. The villages and reindeer roundups in Sattasniemi and Oraniemi cooperatives.

Sources: The map contains data from the Finnish Environmental Institute database (TOKAT database), National Land Survey of Finland (Topographic database 2019), and GADM (database 2015). Drawing Outi Manninen 2021.

south of the Porttipahta artificial lake. It is one of the few remaining *perkka* corrals, which were built out of timber without nails by notching and joining the logs in the corners. A section of the Salmuri corral has been preserved as locally important cultural environment.[82] In the late 1970s it was considered that it should no longer be repaired but be left to decay. However, the majority of the cooperative's members were in favour of mending the corral.[83] Kautoselkä

82. Lapin liitto 2008, Pohjois-Lapin maakuntakaavaselostus.

83. Sattasniemi cooperative general meeting 24 May 1978 §5. Sattasniemi cooperative archives, Ca:7. NA-Oulu. Salmuri round-up was repaired at least in 1909, 1913, 1915, 1950, 1962 and 1978.

roundup site is the other main corral in Sattasniemi cooperative. In the 2010s the local windfarm affected the use of Kautoselkä roundup by disturbing the transportation of the reindeer docks. Forestry has also been said to decrease the opportunities for reindeer nutrition in the area.[84]

The highlights of the reindeer year, roundups, happen in late autumn. The reindeer herders of the research area usually take part in dozens of roundups a year. Information about the place and time was in earlier times announced in the local church,[85] later in the newspaper[86] and nowadays for example by text message.[87] Reindeer owners, their families, buyers and interested locals arrive from all directions – earlier skiing or on reindeer sledges, nowadays with snowmobiles, quad bikes and all-terrain vehicles – and gather at the roundup sites (Map 3), stepping out of the routine of everyday village life.

Technical working tools such as snowmobiles as well as human interventions, such as forestry, mines, roads and snowmobile routes, have changed the logistics system of roundup sites. People travel to corrals with motorised vehicles, which changes the trade in reindeer meat and other products. The first snowmobile was adopted in Sattasniemi in 1964.[88] Both the Sattasniemi and Oraniemi cooperatives used snowmobiles at least from 1965 onwards to drive away carnivores.[89] The popularity of the new tool was not unanimous: in 1966, a member of Sattasniemi cooperative suggested that the use of snowmobiles in reindeer work should be banned, but the meeting decided that snowmobiles could be used when needed.[90] One reason for the dislike might have been the drastic change in the atmosphere and the soundscape snowmobiles caused to

---

84. See Metsahallitus: Kuolavaara–Keulakkopään tuulipuisto. Ympäristövaikutusten arviointiselostus 27 Jan. 2011.

85. See, for example, Kuulutukset (Public notes) 1919, Sattasniemi cooperative archives, Db:2. NA-Oulu.

86. See, for example, the cooperative's meetings on 25 Sept. 1948 §31 and 25 Sept. 1950 §22. Sattasniemi cooperative archives, Ca:5. NA-Oulu; Cooperative meeting 28 May 1983 §16. Sattasniemi cooperative archives, Ca:7. NA-Oulu.

87. Helle 2015, pp. 55, 79.

88. Sattasniemi spring meeting 4 May 1964 §5. Sattasniemi cooperative archives, NA-Oulu.

89. Reindeer tending season 1965–66, Sattasniemi and Oraniemi. Finnish Reindeer Herders' Association archives, Da:2. NA-Oulu.

90. Sattasniemi cooperative extraordinary meeting 30 Dec. 1966 §7. Sattasniemi cooperative archives, Ca:6. NA-Oulu.

forest work, and the motorised vehicles were also said to disturb the reindeer.[91] The old herding style ended in the 1960s due to snowmobiles and, as a result, the reindeer often wandered to neighbouring cooperatives.[92] Radiotelephones were introduced in the area in 1966, when they were advertised as being light and enabling communication.[93] More recently, cellphones have been used to help with the *etto* work. Due to modern tools, reindeer are nowadays easier to locate: they may be tracked with GPS collars and monitored on an iPad.[94]

Figure 1. The reindeer pasturage cycle.

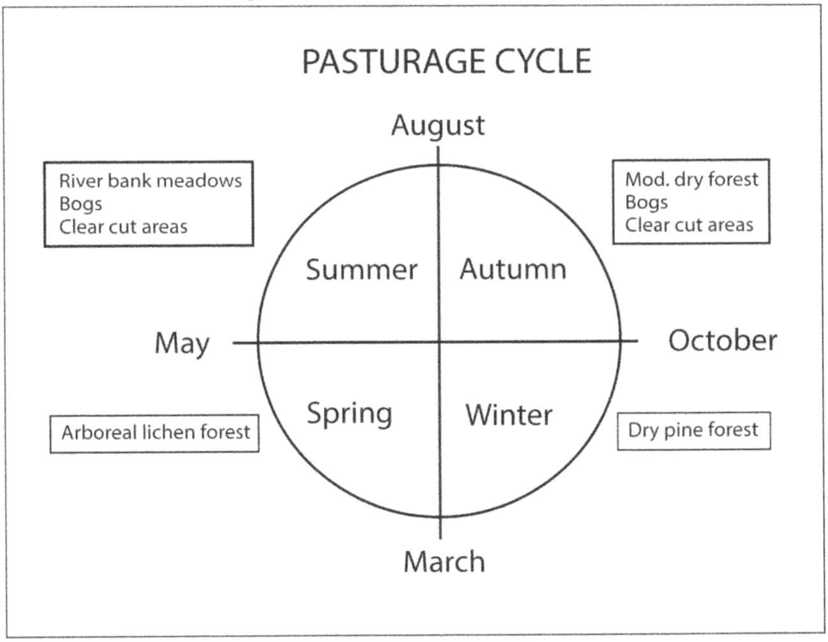

Source: Colpaert et al. 1995, with permission.

91. Sattasniemi cooperative extraordinary meeting 6 May 1974 §1. Sattasniemi cooperative archives, Ca:6. NA-Oulu.
92. Kumpula and Nieminen 1987, pp. 26–27.
93. Reindeer Herders' Association general meeting 1–2 June 1966 §7–8. Finnish Reindeer Herders' Association archives, Da:2. NA-Oulu.
94. Video of Reindeer round-up in Routusvaara, Orajärvi. Seitasäätiö. Mikko Maijala, a member of Oraniemi cooperative, commented on the round-up on 11 Oct. 2016. Nature of the North: Reindeer. See Visit Sodankylä.

Despite the changes in the gathering due to the use of motorised and digital tools, such as GPS collars, helicopters and drones, the process is essentially similar to the one of decades ago. The timetable is still flexible: the period and place of each event depends on the number of animals, weather conditions and other uncontrollable circumstances. Slaughtering, for instance, was for a long time dependent on icy weather, because the animals were slaughtered at the corral: the frost was a natural refrigerator for reindeer meat.[95]

The roundup process starts with gathering the animals in a process called *etto*. Each cooperative decides independently when to go to the forest. Until the mid-1960s, the *etto* took weeks, even months. Spending long periods in the wilderness on foot, on reindeer-sleighs or skiing gradually changed with the adoption of different motorised vehicles. The *etto* process is seen among the herding community as a testimony to the close human-nature relationship. Locals have been assumed to instinctively know how to survive in the wilderness. Moving in the wilderness has been based on traditional knowledge of nature, animal behaviour and weather.[96] Stories in the herders' magazine *Poromies* (Finn. *Reindeer Man,* meaning Herder) from the 1930s paint a romanticised picture of a time, where animals were tame and men skilful. The herders were expected to enjoy their time in nature, not to complain if they had to spend even their Christmas in the snowy halls of the wilderness.[97] The traditional hut (Finn. *kota*) was seen as the only correct way to overnight, and wooden lodges were described as unhealthy and untidy.[98] In a long, romanticised story about an old man visiting his summer pasture hut for the last time, this view can be seen particularly well: nomadic blood is said to run in the old man's veins, which is why he does not want to spend time indoors at all.[99]

The sorting of reindeer gathers a large audience; the roundups have been compared to a trade fair[100] and a 'fiesta of the people of the wilderness'.[101] Even though moving and transportation has become easier, some people still have

---

95. For example, in November 1936 many round-ups had to be cancelled because of the 'prevailing summer weather'. *Poromies* 5–6/1936, pp. 116–19.

96. Ruotsala 2002, p. 325.

97. 'Poromiesporinoita', *Poromies* 11–12/1936, p. 119.

98. 'Asumukset erotusaidoilla kotamaisiksi', *Poromies* 9/1937, pp. 73–74; K.J.H(olster), 'Miten ja minkälaisille paikoille porojen kesäpaimennusta ja merkintää varten olisi porokaarteet eli aitaukset tehtävä?' *Poromies* 6/1931, p. 10.

99. K.J.H(olster), 'Olli-vaarin mietteitä', *Poromies* 6/1935, p. 52

100. 'Poronhoito', *Poromies* 1/1936, p. 5.

101. T. Mäki, 'Teurastuskurssit Oulussa', *Poromies* 6/1939, pp. 102–04.

to spend the night at the roundup sites. Most of the corrals have cabins built around them, and overnighting is still linked to roundups. In a description dating from 1912, the roundup process is said to be chaotic, but fascinating: 'Everything is one big mess.'[102]

Reindeer roundup in Sodankylä in 1962. Photo: Finland National Heritage Agency, public domain.

The roundups are both work and social interaction. Most of our respondents brought up the importance of social life at the corrals and found it essential to bring their families to the roundups to pass on the tradition: 86 per cent of our respondents said their family members participate in the roundup events.[103] One respondent underlined the importance of social interaction; the locals could exchange news, transmit culture from generation to generation and cherish traditions.[104] A modern herder woman normally brings infants to the roundups and forest in a carrier backpack, and the children are used to

102. Heikel et al. 1914, pp. 37–38.

103. Cultural Nature Survey 2021.

104. Cultural Nature Survey 2021: Respondent 11, female, born 1978.

working in nature from an early age, as children of past Sámi families did.[105] The intergenerational aspect of the roundups is strong, which is seen in our survey. One male herder, born in the early 1950s, wrote that his most unforgettable roundup experience was when his grandfather lifted him onto the back of reindeer at the Ylivaara roundup corral when he was four years old.[106] The corral doesn't exist anymore, but it lives in local memories:

> I attended my first roundup in 1956 as a 5-year-old. The same year I had gotten my grandmother's reindeer ear tag. The roundup was held in the Rajala village in Sodankylä at the Ylivaara corral. There were over a thousand reindeer. Many people were dressed in Lappish costumes. … Today when I drive by the Ylivaara roundup corral I see the already partly decayed and fallen corrals. But memories are warm of the past times.[107]

Strengthening of outside regulation

The current strong perception of our respondents is that, as a livelihood, reindeer herding is based on decades-old customs, tacit knowledge and the rhythm of nature, but it also has a growing dimension of regulation and standardisation. Over the last hundred years, big alterations have taken place in the reindeer husbandry. Alongside the transformation toward a more widespread production chain of reindeer meat, the number of reindeer has kept growing and the animals are not as tame as earlier. They have to be forced to the corrals, which has also altered the structure of the roundup corrals.[108]

The communities in Sodankylä together defined the first guidelines for reindeer herding in 1881.[109] Sattasniemi cooperative collected its own guidelines for reindeer care in 1902.[110] This was a sign of putting oral practices on paper and requiring a certain conformity in working habits. One of the writers of the guidelines was Otto Moberg, a reindeer tending adviser[111] and a spokesman of the Finnish Animal Welfare Association (founded in 1901). In winter

105. YouTube. #vaihtovuosisodankylässä vlog 28 Feb. 2020; Interview with Iida Melamies 7 May 2021. See Finnish public broadcasting company.

106. Cultural Nature Survey 2021: Respondent 2, male, born 1952.

107. Cultural Nature Survey 2021: Respondent 4, male, born 1950.

108. Korhonen 2008, p. 18; Kortesalmi 2008, p. 223.

109. Kumpula and Nieminen 1987, p. 26.

110. Sattasniemi cooperative meeting 13 July 1902 §2. Sattasniemi cooperative archives, Ca:2. NA-Oulu.

111. 'Maanviljelijä Otto Moberg 60-vuotias'. *Lapin Kansa*, 10 Sept. 1929.

1913–1914, Moberg visited the roundups in all cooperatives in Sodankylä municipality. He criticised many old habits and stated that animal torture and inconsiderate treatment of the reindeer was still seen in roundups. As an example, he mentioned cases where reindeer were pulled to the corrals, and an untamed reindeer tied to the corral with a leash started to run around. As a result, it was often injured or even decapitated, and the people around might cheer and frighten the animal. Moberg also criticised poor and unhygienic food conditions, which partly led to the deaths of the weakest reindeer.[112]

By the early twentieth century, the structural changes in society and the difficulties of the industry pushed reindeer herders toward wider regulation and cooperation. One target of the reindeer refinement association (Finn. *Poronjalostusyhdistys*, est. in 1926), the predecessor of the Reindeer Herders Association (Finn. *Paliskuntain yhdistys*), was to root out harmful customs in order to improve profitability. According to its rules, the association aimed to promote 'absolutely humane' slaughtering customs and prevent reindeer deaths by bleeding and blood-poisoning from winter markings and castration performed by biting.[113] First, 'the reindeer herders should oblige themselves to handle the reindeer humanely before butchering, without inappropriate pulling and throwing'.[114] More specifically, herders need to treat the animals as fellow mammals. All in all, the animals were to be treated calmly and silently, without unnecessary running, shouting and scaring the reindeer. The timid reindeer were said to fear the corrals and marking made the animals nervous.[115] In 1958 these warning words were repeated; for many, reindeer herding had become an ancillary activity, so reindeer had grown wilder, had little experience of people and therefore were scared of them.[116]

Rapid changes in perceptions of animals has also created local resistance. It feels unchanging, but, in reality, humans' relationship with other animals has always been in transition, and there has always been criticism of how we treat

112. Otto Moberg, 'Kertomus allekirjoittaneen, Suomen eläinsuojelusyhdistyksen asiamiehen, toiminnasta poromailla Sodankylässä talvella vuonna 1913–1914'. *Eläinten ystävä* 10–11/1914, pp. 186–187.

113. Yrjö Alaruikka, 'Suomen Poronjalostusyhdistys 1926–1936', *Poromies* 3/1937, p. 24.

114. Joh. Henr. Herva, 'Porotalouden perus, miten on rakennettava', *Poromies* 1/1931, p. 4.

115. K.J. Holster, 'Sananen porojen kesämerkinnästä ja paimennuksesta sekä porojen kesyynnyttämisestä', *Poromies* 1/1931, pp. 8–9.

116. Reindeer Herders' Association annual meeting 9–10 June 1958 §14. Reindeer Herders' Association archives, Da:1. NA-Oulu.

other living creatures.[117] An animal welfare adviser wrote from Sodankylä in the 1910s that local reindeer herders were willing to take his advice on treating the animals with mercy, but at the same time had a hard time 'giving up the old deep-rooted habits'. As one the most glaring grievances, he mentioned the winter ear markings, because in cold weather the blood would not coagulate, leading sometimes to hemorrhage. He also hoped that the cooperatives would purchase firearms for slaughtering.[118]

The Finnish Animal Welfare Association was trying to get reindeer classified as a domesticated animal, so that their slaughtering would be controlled by law. This goal was achieved in the Animal Welfare Act in 1934, although local veterinarians interpreted based on the 1902 Slaughtering Act that reindeer were private property and should be considered as domesticated animals.[119] Under the Act, the reindeer were supposed to be properly stunned before killing, but it took a long time for this to be done in practice; the reindeer did not necessarily lose consciousness by being hit in the neck.[120] The traditional method was to stab the reindeer in the heart while it was still conscious; the blood was thought to run out so that the meat would be clean.[121] The new methods were encouraged by courses and competitions, for example in slaughtering,[122] with prizes at the roundups for the 'best slaughterers'.[123] Slaughtering at the corrals was evaluated as good, mediocre or bad.[124] There was interest in new slaughtering techniques among the herders, although the old ways were also

---

117. Latva and Lähdesmäki 2020, pp. 19–20; DeMello 2021, p. 19; Bourke 2011, pp. 5–6.

118. 'Rovaniemen paikallisyhdistyksen ja pääyhdistyksen yhteinen poronhoidonneuvoja kirjoittaa Sodankylästä tammikuun 19 p:nä', *Eläinten ystävä* 2/1913, p. 51.

119. They stated that the slaughtering regulations should be announced at churches and supervised by the police. 'Porojen teurastus. Kiertokirje nimismiehille'. Eläinsuojelus 1 March 1912, pp. 41–42; Uutisia: Rovaniemeltä kirjoittaa meille K. Riihiaho seuraavaa. *Eläinsuojelus* 1 March 1912, p. 47.

120. T. Mäki, 'Teurastuskurssit Oulussa', *Poromies* 6/1939, p. 102.

121. Hilda Allén, 'Mikä velvoittaa meitä työskentelemään heikkojen hyväksi', *Eläinsuojelus* 1/1901, pp. 1–65; 'Kauheaa eläinrääkkäystä', *Eläinten ystävä* 3/1907, pp. 49–50; Väinö Virtanen, 'Enemmän inhimillisyyttä poronhoidossa', *Eläinten ystävä* 1/1913, pp. 18–19; 'Porojen teurastus', *Eläinsuojelus* 1 March. 1912, p. 4; T. Mäki, 'Teurastuskurssit Oulussa', *Poromies* 6/1939, p. 102.

122. 'Uutisia', *Eläinten ystävä* 3/1909, p. 49.

123. 'Luettelo hyvistä poronteurastajista', *Poromies* 1/1937, pp. 15–16.

124. See, for example, 'Teurastusneuvojain tiedonantoja marras-joulukuulta 1936', *Poromies* 1/1937, pp. 6–8.

defended.[125] The dislike of outside control is captured in a document about including reindeer in the regulations concerning domestic animal slaughtering. The writer finds it amusing that the legislators talk about 'tame' reindeer, and states that these southern animal welfare utopians have probably seen reindeer only in pictures.[126] Slaughtering at the corral where the other animals could see what was happening was found to be stressful for the animals.[127] In 1955 Sattasniemi cooperative decided to buy a slaughtering pistol after a talk from a reindeer herding adviser (also a member of the cooperative) on how to slaughter reindeer according to the law.[128] In a general meeting of herders in 1958 it was stated that the slaughtering place should be picked so that the other animals would not see the event, and the animal should be properly stunned before killing.[129]

The number of reindeer has risen over the decades, leading to a rise in slaughtering numbers. The slaughtering of calves started in the late 1960s.[130] There is a strict percentage that each owner must mark for slaughter in the roundups, and the decisions are made on an economic basis. Emotions can, however, interfere with rational decision-making: sometimes reindeer owners decide to save a doe already marked for slaughter. This is called 'getting angel's eyes'. The saved individual is often a reindeer that has lived in a pound close to the owner.[131] In the last fifty years, supplementary feeding and keeping the reindeer near the homes of herders have increased contact with the reindeer in the wintertime; herders see them more often, and they might start to resemble a domesticated animal. When an animal gets a name, it is personalised. For example, pulling and racing reindeer have always had names, and they have been seen as valuable individuals. Reindeer racing started in Finland in the

---

125. T. Mäki, 'Teurastuskurssit Oulussa', *Poromies* 6/1939, p. 103.

126. 'Eläinsuojelua koskeva lakiesitys eduskunnalle', *Lapin Kansa* 12 Sept. 1933.

127. 'Mitä tekeillä olevassa porolaissa olisi otettava huomioon', *Eläinten ystävä* 2/1923, p. 17; Frans K. Rantanen, 'Poro, Lapin juhta', *Eläinten ystävä* 5/1934, p. 88.

128. Sattasniemi cooperative meeting 30 April 1955 §15 and §16. Sattasniemi cooperative archives, Ca:5. NA-Oulu.

129. Yearly meeting of the Reindeer Herding Association 9–10 June 1958 §12. Reindeer Herders' Association archives, Da:1. NA-Oulu.

130. Annual reports of Sattasniemi and Oraniemi cooperatives. Finnish Reindeer Herders' Association archives NA-Oulu; Kumpula and Nieminen 1987, p. 29.

131. Helle 2015, pp. 71, 73.

1920s, and the racers got names based on their appearance or skills, such as *Fire hoof*, *Ringed-eye*, *Black* or *Blaze*.[132]

Roughness in roundups, especially in earlier times, might have been related to giving up the animals and hiding personal emotions, as the animals are very much part of the human culture, cultural nature and local self-understanding. In the roundups, some animals were slaughtered and others ran free. In a film of the Sattasvaara roundup in 2020, a female herder, who is also a mother, talks about the sadness she feels when the calves are taken to be slaughtered, and the cows cry for their calves and search for them for days. 'It is quite tough for a female herder', she says, but 'you have to sell the meat to make a living'.[133] Women have gradually been welcomed to participate at the roundup processes as equal reindeer handlers. Traditionally women have been a minority of reindeer herders and owners: in Oraniemi in mid-1980s some nine per cent of reindeer owners were women,[134] but their share has increased in recent decades.[135] This seems to have changed the roundup events to some extent. The change in roundups can be observed, for instance, from the videos filmed in Sodankylä in the 1980s and 2020s. In the older film, the churn is full of men socialising; talking and swearing; some of them are drunk.[136] In a roundup event filmed in 2020 the focus is on female herders. Women, men and children work together, there are far fewer people in the churn, and the atmosphere is structured and peaceful.[137]

---

132. 'Kuusitoista vuotta porourheilua', *Poromies* 1/1938, pp. 7–8.

133. YouTube. #vaihtovuosisodankylässä vlog 28 Feb. 2020..

134. Kumpula and Nieminen 1987, p. 30.

135. A woman was first chosen to be leader of the roundup process in Sattasniemi in 1976 (Autumn meeting 28 Sept. 1976 §19. Sattasniemi cooperative archives, Ca:6. NA-Oulu); The same woman started as the cooperative's secretary in Autumn 1980, appearing for the first time in the board meeting of 12 Sept. 1980. The first woman to lead a co-operative was chosen in the Kaldoavi cooperative in Utsjoki in 2003 (see Siina Välimaa, 'Suomen poroisäntäkin nyt nainen', *Kaleva* 6 June 2003). The first female executive manager of the national Reindeer Herders' Association came in 2010. One implication of the increasing interest in professional herding among women is the increasing number of female students specialising in northern agriculture and reindeer herding: see Lapland University of Applied Sciences and The Sámi Education Institute.

136. Roundup in the Kommattivaara corral (Oraniemi) 1982; Roundup in Sodankylä 1988. In the latter, one individual woman can be seen in the churn.

137. YouTube. #vaihtovuosisodankylässä vlog 28 Feb. 2020.

Another major question concerns protection of large carnivores and other predators, such as the wolf, wolverine, golden eagle, fox, lynx and bear.[138] For a long time, cooperatives granted rewards to those who managed to kill large carnivores.[139] When a law meant to protect bears during the wintertime was discussed in Sattasniemi in May 1964, it raised strong concerns among the herders.[140] In 1966, the chair of Oraniemi cooperative held a talk at the meeting of the Reindeer Herders' Association about bear sightings. The speech, suggesting that the bear protection law should be terminated, got the unanimous support of the participants. Oraniemi's chair complained that the bear had been labelled as an 'innocent herbivore', although it was known to kill reindeer calves. The need for the local circumstances to be acknowledged resonated in the remark that bears should be moved close to Helsinki, so that the people there would get to know the habits of 'the king of our forests'.[141] In the national reindeer herding area, it is still permitted to kill wolves with special permission, but wolverines have been protected from the 1980s onwards.[142]

Conclusions

In the course of the twentieth century, Mid-Lapland has faced enormous environmental change. Intensive forestry, energy production and the mining industry have physically altered the landscape and disturbed reindeer herding based on natural pasture rotation. Continuity of the livelihood and way of life is a worrying issue in the region. Local peoples' feelings of not being heard or understood affect their relationships with nature and reindeer.

Concern about losing the connection to nature and animals was first discussed a hundred years ago, when reindeer herding in Lapland was in transition due to the state-led regulations after 1898, and intensified forestry took space from reindeer herding as a livelihood. Definite modernisation and

138. Cultural Nature Survey 2021: Respondent 7, female, born 1986.

139. For example, in February 1905, it was decided to hand over a male calf for each killed wolf, and in 1909 a female calf for every killed wolf or wolverine. Cooperative meetings 7 March 1905 §19 and 13e Feb. 1909 §9. Sattasniemi cooperative archives, Ca:1. NA-Oulu.

140. Sattasniemi cooperative Spring meeting 4 May 1964, §14–15. Sattasniemi cooperative archives, Ca:6. NA-Oulu.

141. A statement by Aimo Maijala. Reindeer Herders' Association annual meeting, 1–2 June 1966 §11–12. Reindeer Herders' Association archives. Da:2. NA-Oulu.

142. Tuija Sorjanen, 'Porot pysäyttävät sudet Lappiin – "Poronhoitoalue on sudella läpikäymätön raja-aita"', *Lapin Kansa* 5 Jan. 2017.

standardisation of reindeer herding began in the 1930s.[143] This led to a decline in old herding methods, such as shepherding, but also in old habits, such as spending long periods in the pastures and nature.

The more the surrounding natural and cultural environments have changed, the more the Mid-Lappish communities have tried to revitalise the 'original' nature-human-reindeer relationship with nostalgic stories about dense forests, free waterways and untouched wilderness. The locals emphasise their 'authentic' Lappish lifestyle at least in terms of reindeer herding. This endeavour can be regarded as a cultural use of nature. The reindeer corrals have an important role in this cultural survival process. The old corrals connect the participants to the mystical ancient roundup events, and current herders recall repeatedly that their ancestors have worked in the very same places and sites for centuries. According to locals, nature at the roundup corrals, with its tall trees and other vegetation, is constant proof of the long service the corrals have given their home community.

The handling of the animals has changed during the last century, at least in roundup events – becoming more controlled and structured. The phases of the process are essentially the same from *etto* to sorting, but transportation, communication, treatment and slaughtering have evolved. The animals are easier to locate and reach due to motorisation and internet connections, and the time spent in nature has continuously shortened, but still the work is done outdoors.

The way the animals' behaviour and the rhythm of nature are most visibly transferred into human culture is the reindeer year built on the cycles of nature (Figure 1). This feature is one of the major links to locals' so-called Forest Sámi way of life. Mid-Lappish herders have always moved a lot in nature even though they have permanent dwelling places. Natural pasture rotation remains the basis of work in the Sattasniemi and Oraniemi cooperatives.

Our last conclusion concerns gender: reindeer herding and its sub-events are nowadays less masculine, with female herders having gained a significant role in herding and events organised in reindeer corrals. This transition has happened during the last twenty years in the Mid-Lappish communities. While photos and text from the 1970s still mainly show male herders working in the roundups, today they present young female herders who actively post on their social media about reindeer herding, nature and local culture. Owing to these young women's presentation of their culture, reindeer herding sites nowadays are located not only in the physical environment, but in imaginary nature as well.

143. The Reindeer Breeding Association started to revive the livelihood, for example by education and counselling.

Bibliography

Cultural Nature Survey 2021

Conducted by the authors 22 February–30 March 2021.

Archival sources

National Archives of Finland, Oulu

Reindeer Breeding Association archives

Annual reports of the cooperatives 1929–1945

Board and general meeting minutes 1934–1955

The report of the Reindeer Herding Act committee 1932

Finnish Reindeer Herders' Association archives

Annual reports of the cooperatives 1953–1979

Annual reports of the Reindeer Herders' Association 1949–1980

Letters (kirjetoisteet) 1948–1954

Sattasniemi cooperative archives

Public notices 1919

Cooperative meeting minutes 1899–1986

Board minutes 1968–1986

Official letters 1911–1944

Private letters 1903–1942

Sodankylä Reindeer Herding Association archives

Cooperative minutes and extracts 1936–1948

Magazines and newspapers:

Eläinten ystävä 1907–1934

Eläinsuojelus 1901, 1912

Kaleva 2003

Lapin Kansa 1928–1939, 2010–2020

Metsästys ja kalastus 1937

Pohjolan Sanomat 1935–1939

Poromies 1931–1939, 2010–2020

Rovaniemi 1934–1939

Sompio 2010–2020

Suomen Luonto 1959

Online sources

Arndt, Melanie, Environmental History 2016: https://docupedia.de/zg/Arndt_environmental_history_v3_en_2016 (accessed 16 June 2021).

Close Reading and Annotating Historical Sources: https://lessonresearch.net/content-resource/close-reading/ (accessed 16 June 2021).

GADM database 2015: https://www.diva-gis.org/ (accessed 30 June 2021).

Finnish Environmental Institute (SYKE), nature protected areas: https://ckan.ymparisto.fi/dataset/luonnonsuojelu-ja-eramaa-alueet (accessed 30 June 2021).

Finnish Environmental Institute (SYKE), TOKAT database: https://www.syke.fi/hankkeet/tokat (accessed 30 June 2021).

Finnish Heritage Agency Database: https://museovirasto.finna.fi/ (accessed 20 May 2021).

Finnish public broadcasting company (Yle),Interview with Iida Melamies: https://yle.fi/aihe/artikkeli/2021/05/07/iida-melamies-haluaa-vaikuttaa-poronhoidon-tulevaisuuteen-ja-nyt-nuori-nainen (accessed 27 May 2021).

Instagram, local reindeer entrepreneurs 2021.

Järvenpää, Juha. Poro ja poronhoito talousmetsissä: Katsaus metsätalouden ja porotalouden yhteensovittamiseen Suomessa 2018: https://paliskunnat.fi/ohjeet_oppaat/Poro_ja_poronhoito_talousmetsissa_2018.pdf (accessed 30 June 2021).

Lapin liitto: *Pohjois-Lapin maakuntakaava.* Inari – Sodankylä – Utsjoki kaavaselostus 2008: https://www.lapinliitto.fi/wp-content/uploads/2020/11/Pohjois-Lapin-maakuntakaavaselostus.pdf (accessed 7 March 2021).

Lapland University of Applied Sciences: https://www.lapinamk.fi/fi/Hakijalle/AMK-tutkinnot/Agrologi,-maaseutuelinkeinot/Opiskelijatarinoita (accessed 25 June 2021).

Metsähallitus, Kuolavaara–Keulakkopään tuulipuisto. Ympäristövaikutusten arviointiselostus 27 Jan 2011: https://www.metsa.fi/wp-content/uploads/2020/06/Kuolavaara-YVA-selostus.pdf (accessed 7 March 2021).

Metsälappalaispäivät ry.: https://www.metsalappalaiset.net/ (accessed 12 April 2021).

National Land Survey of Finland, Topographic database 2019: https://www.maanmittauslaitos.fi/en (accessed 30 June 2021).

Paliskuntain yhdistys, Poron hoito ja käsittelyopas 2009: https://paliskunnat.fi/py/wp-content/uploads/2014/12/poron_hoito_ja_kasittelyopas_2009.pdf (accessed 31 May 2021).

Paliskunnat 2020: https://paliskunnat.fi/poro/ (accessed 15 November 2022).

Piltz Martti, Selvitys Seitaniemen tien museoarvosta 2009: https://www.mobilia.fi/sites/default/files/seitajarvi_selvitys.pdf (accessed 7 March 2021).

Round-up in Sodankylä 1988: https://www.youtube.com/watch?v=9Aa79JXH8pI&t=334s (accessed 25 May 2021).

Sakatti Mining Company / University of Eastern Finland, Porotalousselvitys. Sakatin monimetalliesiintymän kaivoshanke – Poroselvitys, 25 Nov. 2020: https://finland.angloamerican.com/~/media/Files/A/Anglo-American-Group/Finland/environment/sakatin-ymparistovaikutusten-arviointi/liite-20-porotalousselvitys.pdf (accessed 20 May 2021).

Sámi Education Institute: https://www.sogsakk.fi/%2Fen (accessed 25 May 2021).

Sattanen village homepage: https://sattanen.info/wordpress/?page_id=53 (accessed 27 May 2021).

Visit Sodankylä, Video of a Reindeer round-up in Routusvaara, Orajärvi. Seitasäätiö. Nature of the North: Reindeer: https://www.visitsodankyla.fi/en/pohjoisenluonto-artikkeli/porot-2/ (accessed 6 May 2021).

Ylilokka Eila 5 May 2016: https://www.metsalappalaiset.net/kirjoituksia (accessed 12 April 2021).

YouTube, #vaihtovuosisodankylässä vlog 28 Feb. 2020: https://youtu.be/WHhW4OkhACA (accessed 25 May 2021).

YouTube, Reindeer round-up in Orajärvi Routusvaara 11 Oct. 2016: https://www.youtube.com/watch?v=sSdUmQNgR3w (accessed 6 May 2021).

YouTube, Round-up in Kommattivaara corral 1982: https://www.youtube.com/watch?v=Hu-5JQ0M1JhA (accessed 25 May 2021).

Literature

Antikainen, Janne et al. 2019. *Metsälappalainen kulttuuri ja sen edistäminen*. Helsinki: Valtioneuvoston kanslia.

von Bonsdorff, Pauline. 2007. 'Maisema toiminnan ja kuvittelun tilana'. In Yrjö Sepänmaa, Liisa Heikkilä-Palo and Virpi Kaukio (eds). *Maiseman kanssa kasvokkain*. Helsinki: Maahenki. pp. 33–49.

Bourke, Joanna. 2011. *What It Means To Be a Human. Reflections from 1971 to Present.* Berkeley: Counterpoint.

Colpaert, Alfred, Jouko Kumpula and Mauri Nieminen. 1995. 'Remote sensing, a tool for reindeer range land management'. *Polar Record* 31 (177): 235–244.

DeMello, Margo. 2021. *Animals and Society. An Introduction to Human-Animal Studies.* New York: Columbia University Press.

Gadgil, Madhav, Fikret Berkes and Carl Folke. 1993. 'Indigenous knowledge for biodiversity conservation'. *Ambio* 22 (2–3): 151–56.

Heikel, T.A. et al. 1914. *Porolaidunkomisioonin mietintö.* Helsinki.

Heikkinen, Hannu I., 2012. 'Perinteisen karjatalouden julkisivut ja takapihat Arizonassa'. In Timo Kallinen, Anja Nygren and Tuomas Tammisto (eds). *Ympäristö ja kulttuuri.* Helsinki: University of Helsinki. pp. 305–28.

Heikkinen, Hannu I., Simo Sarkki and Mark Nuttall. 2012. 'Users or producers of ecosystem services? A scenario exercise for integrating conservation and reindeer herding in northeast Finland'. *Pastoralism: Research, Policy and Practice* 2 (1): 1–24.

Helle, Timo. 2015. *Porovuosi. Tutkija pororenkinä Sompiossa.* Helsinki: Maahenki Oy.

Hughes, J. Donald. 2008. 'Three dimensions of environmental history'. *Environmental History* 14 (4): 1–12.

Hughes, J. Donald. 2006. *What is Environmental History.* Cambridge: Polity Press.

Isenberg, Andrew C. (ed.) 2014. *The Oxford Handbook of Environmental History.* New York: Oxford University Press.

Korhonen, Teppo. 2008. *Poroerotus – historia, toiminta ja tekniset ratkaisut.* Helsinki: Finnish Literature Society.

Kortesalmi, J. Juhani. 2008. *Poronhoidon synty ja kehitys Suomessa.* Helsinki: Finnish Literature Society.

Koster, Egle, Kajar Koster, Mika Aurela, Tuomas Laurila, Frank Berninger, Annalea Lohila and Jukka Pumpanen. 2013. 'Impact of reindeer herding on vegetation biomass and soil carbon content: a case study from Sodankylä, Finland'. *Boreal Environment Research* **18** (6): 35.

Kumpula, Jouko, Mika Kurkilahti, Timo Helle and Alfred Colpaert. 2014. 'Both reindeer management and several other land use factors explain the reduction in ground lichens (Cladonia spp.) in pastures grazed by semi-domesticated reindeer in Finland'. *Regional Environmental Change* **14** (2): 541–59.

Kumpula, Jouko and Mauri Nieminen. 1987. 'Oraniemen paliskunnan poronhoidosta, porolaitumista ja porotaloudesta'. *Poromies* 1/1987: 24–31.

Laurén, Kirsi. 2006. *Suo – sisulla ja sydämellä. Suomalaisten suokokemukset ja -kertomukset kulttuurisen luontosuhteen ilmentäjinä.* Helsinki: Finnish Literature Society.

Latva, Otto and Heta Lähdesmäki. 2020. 'Miten kertoa menneisyydestä ja rakentaa tulevaisuutta – historiasta, ihmisistä ja muista eläimistä'. In Elisa Aaltola and Birgitta Wahlberg (eds). *Me & muut eläimet. Uusi maailmanjärjestys.* Tampere: Vastapaino.

Little, Paul E. 1999. 'Environment and environmentalism in anthropological research: Facing a new millennium'. *Annual Review of Anthropology* **28**: 253–84.

Lähteenmäki, Maria. 2006a. 'Kemin Lapin raunioilla'. In Maria Lähteenmäki (ed.) *Alueiden Lappi.* Rovaniemi: Lapland University Press. pp. 60–79.

Lähteenmäki, Maria. 2006b. *The Peoples of Lapland. Boundary Demarcations and Interaction in the North Calotte from 1808–1889.* Helsinki: Finnish Academy of Science and Letters

Lähteenmäki, Maria, Oona Ilmolahti and Alfred Colpaert. 2019. 'Nature represented. Environmental dialogue in the Finnish-Karelian museums'. *Museum International* **71** (2): 88–105.

Merchant, Carolyn. 1987. 'The theoretical structure of ecological revolutions'. *Environmental Review* **11** (4): 265–74.

McNeill, John R. 2003. 'Observations on the Nature and Culture of environmental history'. *History and Theory* (4): 5–43.

Pilgrim, Sarah and Jules Pretty. 2010. *Nature and Culture: Rebuilding Lost Connections.* London Routledge.

Potinkara, Nika. 2012. 'Pohjoisen jalot villit? Alkuperäiskansan maaoikeudet ja luontosuhde etnopoliittisessa keskustelussa'. In Timo Kallinen, Anja Nygren and Tuomas Tammisto (eds). *Ympäristö ja kulttuuri.* Helsinki: University of Helsinki. pp. 281–302.

Pyhäjärvi, Leena. 2011. 'Lokka muutosten näyttämönä – allasalueen elinkeinojen muutos'. In Leena Pyhäjärvi et al. (eds). *Lokka muutosten näyttämönä.* Rovaniemi: Lapin tutkimusseura ry.

Rannikko, Pertti. 2020. 'Suden salametsästys ja vaikenemisen kulttuuri'. *Oikeus* **49** (1): 74–93.

Richardson, Michael. 2001. *The Experience of Culture*. London: Sage.

Ruotsala, Helena. 2002. *Muuttuvat palkiset. Elo, työ ja ympäristö Kittilän Kyrön paliskunnassa ja Kuolan Luujärven poronhoitokollektiiveissa vuosina 1930–1995*. Helsinki: Suomen Muinaismuistoyhdistys.

Sarivaara, Erika and Satu Uusiautti. 2013. 'Taking care of the ancestral language: The language revitalisation of non-status Sámi in Finnish Lapland'. *International Journal of Critical Indigenous Studies* 6 (1) 2013: 1–16.

Smith, Laurajane. 2006. *Uses of Heritage*. London: Routledge

Stark, Sari, Outi M. Manninen, Oona Ilmolahti and Maria Lähteenmäki. 2022. 'Historical reindeer corrals as portraits of human-nature relationships in Northern Finland'. *Arctic Journal* 75 (3): 330–43.

Strand, Marita. 1994. *Saarivaaran vanha poroaita Savukosken Tanhuassa*. Helsinki: Museovirasto.

Williams, Raymond. 1980. 'Ideas of Nature'. In *Problems in Materialism and Culture: Selected Essays*. London: Verso. pp. 67–85.

Chapter 7

SÁMI FRAMES IN THE PLANNING AND MANAGEMENT OF NATURE PROTECTION AREAS IN HISTORICAL PERSPECTIVE – ENVIRONMENTAL NON-CONFLICT IN INARI

Jukka Nyyssönen

Introduction

Northernmost Lapland is well conserved. In addition to numerous protected areas (PAs),[1] nine of the twelve wilderness areas are located in Upper-Lapland, where ninety per cent of the land area and 53 per cent of forest lands are protected.[2] This article concentrates on Inari due to the uniqueness of its timberline forests, an ecotone between the northern boreal forest zone and the tundra.[3] Inari is a multi-ethnic municipality, populated by Finns and three Sámi[4] groups, distinguished, for example, by their languages, Northern, Aanaar and Skolt Sámi. In Finland, the Sámi enjoy constitutional self-government and the status of an Indigenous People (IP).[5] Another distinctive feature in Upper-Lapland

1. Nature protection terminology is abundant. This article covers mostly cases of national parks (kansallispuisto), open for different usage and tourism. When relevant, I use the specific term for each protected area. As a general term, I use the term Protected Area (PA). On terms, see e.g. Koilliskairatoimikunta 1972, p. 3.
2. Situation in 2006. Upper Lapland includes the municipalities of Enontekiö, Utsjoki and Inari. Hallikainen et al 2008, pp. 192–93, 203.
3. Raitio 2008, p. 81.
4. The Sámi reside in Finland, Sweden, Norway and Russia. The current total population is estimated to be appr. 70,000–100,000. The Sámi are the only folk enjoying the status of Indigenous people within the EU. The Sámi speak 9 surviving languages and practise versatile sources of subsistence. A significant minority is engaged in reindeer herding.
5. www.samediggi.fi (accessed 17 Oct. 2022).

doi: 10.3197/63824846758018.ch07

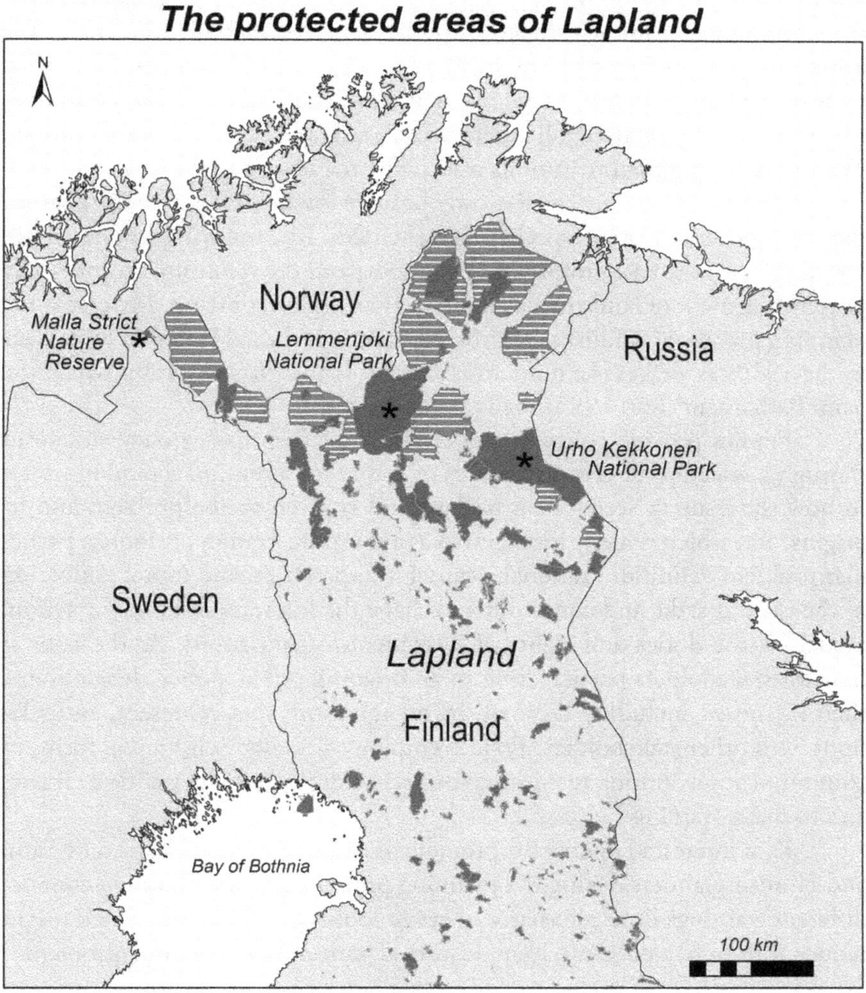

Map 1. Much of the protected area is located in Upper Lapland. Not even in these areas is conservation total, as in the Wilderness areas (hatched red) light forms of multiple use are allowed. Map by Jan Magne Gjerde using ArcGIS and data from the Finnish Environment Institute (accessed Feb. 2023), CC 4.0.

is recurring and long-lasting disputes over resource and land-use, which have centred in the municipality of Inari. The main reason for disputes has been the relationship between reindeer herding and forestry. The fronts have been re-formed many times, the conflicts do not follow ethnic boundaries, and actors involved range from local to global. Another source for disputes has been the seemingly continuous stream of proposals to establish PAs in Inari. The initiatives have provoked severe resistance against nature conservation, caused by fears over loss of lands, usage rights and raw materials; and the perception of undemocratic dictated policies from the south and neglect of local opinion. The EU is among the institutions resisted by the local population.[6]

One group has remained mostly positive towards PAs up to the present: the Sámi and among them especially Sámi herders.[7] By employing frame analysis, I shed light on the background of this support and the resulting, exceptionally peaceful segment of land-management in Inari region, Finland. My case is the planning process of Koilliskaira/Urho Kekkonen National Park that took place in the 1970s as well as the numerous cases of nature protection on which the Sámi Parliament[8] (est. 1995) made statements.

Frames provide coherent understanding of complex policy situations through a selection of certain features of reality for attention. Framing relates to how the issue is 'seen', what background is given to the problem and its origins, and which remedy is suggested as a solution. Frames promote a particular problem definition, rationale, causal interpretation and moral evaluation of the issue at stake and condition and shape the interests and bias for action, as well as the duties and rights of individuals/organisations. As the issue is framed, stakeholders participating in an ongoing public policy debate frame their identities, including those of the organisations they represent, and relations with other stakeholders. Frames employ culturally well-known forms of communication, aiming to create resonance with the broader political climate and to make framing successful.[9]

I am interested in how the problem of conservation was framed by Sámi and Finnish planners during the planning processes. In the planning context, different framings limit alternative ways of looking at the issue, which might reduce it to only a political, legal, historical, cultural or economic problem.[10]

6. Hallikainen et al. 2006, pp. 459–62, 464, 467; Veistola 2008, pp. 241–43.

7. Hallikainen et al. 2006, pp. 467–72; Markkula, Turunen and Kantola 2019, passim.

8. Sámi Diggi (est. 1995) is the responsible organ for the cultural autonomy the Sámi received the same year. The Parliament has a right of initiative. According to the Act on the Sámi Parliament (1995), the state authorities shall negotiate with the Sámi Parliament in all far-reaching and important measures which may directly affect the status of the Sámi as an IP and which concern the management, use, leasing and assignment of state lands, conservation areas and wilderness areas. Sara 2019, p. 32.

9. Creed, Langstraat and Scully 2002, p. 37; Raitio 2008, pp. 45–49; Sara 2019, 18ff.

10. Sara 2019, p. 122.

This limiting function of the frames resembles 'whole story frames', interested in what each party believes the dispute is about, which guides their argumentation in the dispute.[11] I use whole story frames to reveal the underlying interests and whether the frames and the policy choices aligned in the negotiation over national parks in Inari.

I argue that Sámi planners framed the issue for a long time as economic. I also argue that the increasingly powerful environmentalist frame did not hinder frame alignment between key actors, since the interests overlapped, while the principal issues of Sámi rights have been introduced independently of the environmental frame. At times, the environmental frame shared a similar definition of the problem, with the Sámi frames. The focus is on how Sámi manoeuvred the administrative landscape and how it was used as a resource by the Sámi (herders).

This article contributes to the field of conservation history, an essential ingredient in the study of environmental history. The genre has focused on projects of building of national identity and on the aesthetic, religious and ethical convictions motivating conservation.[12] This article touches upon the much studied issue of IP and nature protection. Typically, the IP have been represented as possessing a knowledgeable, religious/spiritual and warm relation to their environments. This old representation has numerous potential pitfalls and has made the alliance between environmentalists and the Sámi fragile, because of the expectations of authenticity connected to an environmentally sound relationship to nature. If these expectations have not been met, paternalism and intolerance have surfaced.[13]

The Sámi relation to nature is a complex, multi-layered issue. The ways the Sámi have in different times held aspects of their physical environment holy and built their worldview through it, is well-documented in Sámi research.[14] In addition to carrying the Sámi relation to nature, reindeer herding comprises a whole life-system, creating identity and maintaining social networks. Herding sustains aspects of culture and contains ethical and juridical-political aspects.[15] The sources used in this article, consisting of official publications, committee reports and statements given by the Sámi Parliament on park planning, do not contain information on these issues. They have their origin in administrative

11. Kennedy, Brown and Butler 2021, passim.
12. Niemi 2018, pp. 18–19.
13. Vincent and Neale (ed.) 2016, passim.
14. E.g. Äikäs 2019, passim.
15. Nordin 2008, passim.

organs and contain a number of state-pronounced biases and guiding grids that in part framed the replies herding community gave. The sources do not contain direct information on Sámi perceptions of their lived environment, so this cultural and spiritual aspect of PAs will not be touched upon.[16] The sources reveal the priorities and frames of the Sámi (Sámi herders from the herding administration and members of the Sámi Parliament), who received a say in the planning processes, taking an active role in the park planning processes. In addition, the actors include the organisations and individual key actors responsible for nature protection and initiating and articulating the environmentally inspired frame.[17]

Research on nature protection in Finland has long showed clear preference for natural values.[18] Nowadays the extensive research on PAs and the Sámi has engaged actively in discussion of the Sámi rights connected to conserved areas. Scandinavian studies have long been concerned about the cultural rights of the Sámi to (traditional) usage of natural resources.[19] The demands of biodiversity were not allowed to overrule this aspect.[20] Restrictions on Sámi usage, types of herding and lack of statutory protection and true stakeholder participation are problematised in Scandinavian research, while herders are mostly positive towards PAs. For example, in Norway, herder fears are connected to tourism and development plans in PAs but, in the 2000s, PAs are understood increasingly as sites for generating more comprehensive rights in future and raising principal issues of rights to land, self-determination and true say in PA management and use of nature and natural resources. Not only ecological sustainability, but also socio-cultural and economic sustainability, defined from Sámi premises, must be taken into consideration.[21]

In certain segments of park studies, most notably in inquiries in social sciences and cultural studies, nature conservation is represented as a predatory and colonialist undertaking, where peripheric land of lesser use value is conserved

---

16. Heinämäki, Herrmann and Neumann 2014, p. 192.
17. In the 1970s, a separate office of the inspector of nature protection (Luonnonsuojeluvalvojan toimisto) at the Finnish Forest Research Institute (Metsäntutkimuslaitos) was responsible for the conservation of nature. The management and planning were taken over by luonnonvarainhoitotoimisto at the Ministry of Agriculture and Forestry in 1973. Ahonen 1997; Joutsamo 2008b, p. 130; Suominen 2008, p. 85.
18. E.g. Telkänranta (ed.), 2008, passim.
19. Allard 2017, p. 9; Andreassen 2002, p. 124; Elenius 2017, p. 1.
20. Andreassen 2001, passim; Green 2009, passim; Schanche 2001, pp. 4–7.
21. Andreassen 2002, pp. 115–20; Reimerson 2021, p. 77.

to boost the national identity of the coloniser. Scientific needs and arguments about preservation of biodiversity and ecology have been tools to dispossess and bypass native communities in terms of land ownership rights, since the coupled authority and expertise to talk about these issues is reserved to state authorities. Science and ecology bypass the moral issues of indigenous rights and stewardship over ancestral lands.[22] Another related tendency in recent park management studies is to call for and analyse existing ways of stakeholder inclusion in PA management, based on dialogue, collaborative action, partnership and social learning.[23] Although these studies have revealed numerous histories of dispossession and contributed to conservation studies in settler colonies, and in Sámi areas,[24] their approaches are not followed in this study, since, in the era studied, rather than colonial dispossession the process in Finland was one of (limited) inclusion of the Sámi, which is analysed as it emerges from the sources. As such, the systemic unbalances in the processes are of interest in this research, but they are studied without grids from recent theorising.

The theme of this chapter is of successful managemental encounters between the Sámi and the Finnish administration. The research questions to be illuminated include: How the Sámi framed their interests in the planning processes? Did these frames resonate with competing (forestry, ecological) frames? Did the herders manage to use the administrative setting as a resource, or did it constitute a hindrance to their interests? Which factors in the institutional setting contributed to the planning process?

Inari: History of multiple uses, crossing interests – and conservation

The organ responsible for forestry in state owned forest, the Forest and Park Service (FPS), has a history of working hard to launch and sustain forestry in Inari. This project was hindered by long distances, poor transport connections to the mills in southern Lapland and high costs. The local population, including the Sámi, had been actively involved in the forestry project. The FPS, managing the nature conservation areas as well, constituted a centre of power, with tangible consequences for herders due to usage areas overlapping with reindeer herding. As the efficient forms of forestry were finally introduced to Inari in the 1960s

22. Andersson, Cothran and Kekki 2021, passim; Green 2009, pp. 56–57.

23. Andersson 2019, passim; Andersson, Cothran and Kekki 2021, passim; Getzner, Vik, Brendehaug and Lane 2014, passim.

24. Malla Strict Nature Preserve is given as an example of dispossessing the Indigenous Sámi from their pastures: herding is prohibited, in the name of research and a 'pristine' state of nature. Andersson, Cothran and Kekki 2021, pp. 3–4.

and 1970s, it took a decade before conflicts arose.[25] Ownership of reindeer and right to practise herding are allowed to Finns as well, but the Sámi herders form a majority in the herding cooperatives (*paliskunnat*) of Upper-Lapland.

The northernmost forests of Finland have long been considered special and worthy of conservation, due to their untouched, 'wilderness'-like character. Ecological, scenic and aesthetic values have ranked high, while usage values, varying from touristic use to forestry, are subject to ethical flexibility. Conservation began at the start of the twentieth century with the preservation of natural monuments and landscape protection. The first modern conservation area, meant to protect endangered plants, was established in 1916 in Malla, Enontekiö (nature reserve from 1938 onwards). Numerous laws restricted, but did not totally prevent, logging, and forests were parcelled into several categories. The forests highest on the mountain slopes (*suojametsät* in Finnish), with low economic gain potential, were conserved in 1939.[26]

The FPS-rationale was conservation of distant forests, wishing to minimise the economic loss. Forests with landscapes worth conserving were protected by forester-initiative and by FPS official policy from 1903 onwards (called as *säästömetsät* in Finnish). Early PAs (until 1922 five of them) were established following FPS-initiative and expertise. The protection of the northernmost forests was motivated by fear of loss of forest resources, as the slow and uncertain re-generation of the northern forests dawned on foresters.[27] The FPS lost its exclusive power to preserve forest areas in the planned Law on conservation (*Luonnonsuojelulaki*) in 1923. Protection of endangered animal and plant species was now listed as a reason for conservation/preservation, alongside aesthetic reasons. The law, which was postponed due to the unclear landownership situation in the northernmost Lapland, would have meant real restrictions on land-use, and these limitations to economic utilisation of the areas caused severe resistance to the law.[28]

Reindeer herders were not included in the early use planning of the northernmost forests, but reindeer herding was one of the factors taken into consideration when the FPS and a special committee reserved high-lying forests in the early twentieth century (*Suojametsälaki*, legislation in 1922).[29] Later, individual Sámi, along with all the residents of Inari, had the opportunity to

25. See e.g., Nyyssönen 2000, passim.

26. Joutsamo 2008a, pp. 80–81; Nyyssönen 2000, pp. 141–42.

27. Parpola and Åberg 2009, pp. 123–24.

28. Ibid., pp. 128–29.

29. Ibid., pp. 124–28.

make remarks on park planning at special hearings.[30] Economic framing was typical for local stakeholders: in the 1930s, in the case of the protected area of Heinäsaaret in Petsamo, locals feared that the expansion of seagull populations would make fishing more difficult. They also feared increase in predator numbers as well as restrictions on reindeer herding and agriculture.[31] Economic framing also dominated comments on the planning of Lemmenjoki National Park in the 1950s: herders were already then the most positive about protecting the region, as long as they received the right to hunt predators.[32] I now turn to the planning process of Koilliskaira.

Koilliskaira and the inclusion of herding – economic frame

In 1977, a nominated state organ, a nature protection branch of the advisory committee for environmental protection in the Ministry of Interior (*Ympäristönsuojelun neuvottelukunnan luonnonsuojelujaosto*) aimed to widen the network of PAs. The proposition was extensive: 42 new national parks and sixteen new nature reserves plus an increase in the area of existing national parks and reserves were suggested, totalling 1,161,000 hectares. The report received fierce criticism, especially from private landowners, the lumber industry and local people. In Inari, local FPS officials were negative to the plan and engaged in precautionary measures typical at the time: logging and road construction in fringe areas adjoining and within the planned protected areas. The critique concentrated on the fear of losing use rights, job opportunities and opportunities for rational forestry in Inari. The motivation for protection was linked to the values of nature. The establishment of a reduced number of PAs was postponed until 1982.[33]

The case of Koilliskaira is illustrative of the new thinking on the environment: PAs served a function of preserving nature in its original state. The inherent right of the ecosystem to existence was acknowledged as a starting point for conservation. In addition to recreational and aesthetic values, the general capacity to function as an ecosystem and the preservation of the ecosystem's genetic information were the elements protected. Additional aims included conserving a representative sample of typical ecosystems, all age-classes and

30. Kansallispuistokomitean mietintö 1976, pp. 88, 67, 70.
31. Vahtola 1999, p. 500.
32. Luonnon- ja kansallispuistokomitean mietintö 1953, pp. 9, 38–42.
33. Kansallispuistokomitean mietintö 1976, pp. 37–38; Ahonen 1997.

landscape types for each province, in addition to which conservation took place at all scales, ranging from single endangered species to the global scale.[34]

This was the authoritative, ecological framing of the matter. The planning organs articulated a parallel economic frame. Instead of imagery of untouched nature, the planning of the Koilliskaira took multiple uses of the conserved area as a point of departure. In the 1970s, special concern was laid on the possibilities for local people's continuing subsistence and economic compensation for conservation. Reindeer herding was seen as a natural, thriving part of the preserved areas and ecosystems, as reindeer could benefit from restrictions on land use and motorised traffic in the terrain.[35] Reindeer appeared robust animals, grazing 'almost in all kinds of habitats', turning almost all the protected areas into reindeer pasture.[36] The Sámi were claimants enjoying ecosystem services and benefiting from the protection. The framing is economic, not eco-romantic, arising from Sámi needs, not of imaginings of Sámi immersion in nature.

The Reindeer Herders' association (*Paliskuntain yhdistys*) and the Sámi herders received a voice, as a herder from Sodankylä, Jouni Aikio was invited to the commission. Reindeer herding was defined as a 'central form of economy and life', as an original and innate means of living protected by special legislation in the areas to be protected, as a significant employer and as a land use form that did not threaten the aim of protecting nature in its natural form. The commission referred to the 'strong protection' of herding, not least by national herding and conservation legislation. Reindeer herding qualified as a means of living based on 'nature's economy', falling thus under the recommendations of International Union for Conservation of Nature (IUCN), which took these kinds livelihoods as the best guarantee of nature preservation. That reindeer herding, although capable of causing harm for landowners, was not taken as an ecologically harmful means of living, but as subsistence threatened by externalities, made its inclusion smooth.

All the planners took efficient forestry to be harmful to the winter pastures, especially if soil scarification was employed. Another externality was unregulated tourism, considered at the time as a great nuisance to herding. The commission recommended that protective measures and the status of herding had to be made statutory, in order to improve the protection of herding. Even though the commission suggested cuts to the protected area, trying to balance the demands of securing employment, the most important means of

34. Koilliskairatoimikunta 1972, p. 1.

35. Kansallispuistokomitean mietintö 1976, pp. 19–27, 61, 83–84, 87, 116, 118, 142, 158–61, 170.

36. Koilliskairatoimikunta 1972, pp. 1–44, 52, 112.

protection was to limit forestry to the fringe areas of the park and to limit its efficiency too. Reindeer herding was to be allowed in all the park area, as were hunting and fishing.[37]

A number of herding experts, including Sámi herder Jouni Aikio, forming a working group, were consulted in the planning of Koilliskaira. For the most part their arguments followed that of the 1972 commission, but they stressed more the vulnerable situation of winter pastures. The working group also employed a more sophisticated classification of pasture lands. Knowledge of the effects of forestry on reindeer pastures was more detailed and perception sharper; efficient forestry aggravated the already precarious situation of herding (the national park bordered former pasture areas now under the reservoir of Porttipahta, in the Lappi herding cooperative). Only restricted forms of forestry were to be allowed in the adjoining areas (the area protected was thus considerably larger than the one suggested by the 1972 commission). Herders wished for restrictions on soil processing and for trees with arboreal lichen only slightly logged.[38] These aspects related to damage to the growing capacity of pastures and preserving a winter grazing resource for the reindeer. The framing was economic, concerned about preserving the prerequisites of herding.

The protection of herding meant that the loss of employment, including knock-on effects in local economy, was predicted to be nil. There was no need to compensate for loss of employment, as in the case of forestry. The prognosis of the economic frame,[39] that of low negative economic impact from protection,[40] was in alignment with the herder frame.

One issue that caused different reactions and recommendations among the planners was the question of predators. Herders thought the losses caused by predators were significant and feared that the protection of viable predator populations would risk the economic sustainability of reindeer herding outside the park borders too. The working group had collected negative statements from neighbouring cooperatives and argued in terms of cultural rights: reducing reindeer herding to rearing reindeer as fodder for predators would entail end of herding as a meaningful life-form.[41] This is a rare case of fetching support from another 'whole story'- and identity frame, that of Sámi cultural rights and cultural survival, from outside the process and the dominant economic frame.

37. Koilliskairatoimikunta 1972, pp. 1–44, 52, 112.
38. Selvitys 1972, pp. 9–10.
39. Frames are diagnostic and prognostic. Raitio 2008, pp. 50–51.
40. Koilliskairatoimikunta 1972, pp. 50–52, 59, 74.
41. Selvitys 1972, pp. 9–10.

The working group suggested special legislation to improve the protection of herding: the area should be diminished if this legislation was not put in place. If it was, then even a larger area could be protected. Existing nature protection legislation could not provide sufficient protection for herding. The herding community and its administrative organisations saw nature protection in an instrumental manner, as means to protect their subsistence. The statute suggested by the working group was meant to secure the right to herd and to use herding infrastructure, as well as to hinder too-efficient logging in the protected areas.[42] The framing was economic but could be aligned with the environmental frame concerned about natural values.

The national park committee of 1976 did not wish to voice the industrial discourse about harmful reindeer evident, for example, in forestry research and from the FPS;[43] on the contrary, herding and pastures were in need of protection and protection was in the interest of reindeer herding. The rutting and calving periods were represented as periods of vulnerability to disturbance from tourism and motorised traffic.[44] Adding to this pro-herding stance, the exclusion of local FPS officials from the planning process was one of the most decisive factors for the continuing peace after the planning process.[45] During the process, the economic sustainability of herding frame of the herders and the conservation frame of the central planning organs, committees and environmental actors aligned because of shared mistrust of the FPS. The other risk for Sámi and herder reputation, too great numbers of reindeer, was not (yet) articulated as a risk, as the condition of pastures was generally good and the externalities posed a greater, already observable threat.

While a coalition could be built with the ecological framing, the herders utilised culturally well-known forms of communication and a selection of categories, by seeing herding mostly as a source of economic sustainability. This economic framing of herding resonated well with the official and administrative framings of the era, still echoing the economic imperatives of employment, needs of industry and GNP. The remedy suggested by the herder framing was the same as in the ecological framing: exclusion of forestry. The Koilliskaira planning process is a case of creating, finding and using resonance across state frames, to build a successful front against another powerful state actor, the FPS.

---

42. Ibid.

43. Nyyssönen 2022, passim.

44. Kansallispuistokomitean mietintö 1976: 88, pp. 61, 83–84, 87, 116, 118, 142, 158–61, 170.

45. Nyyssönen 2000, p. 168.

Statutory protection by the bureaucratic state was considered sufficient, which it indeed turned out to be, since the park statutes kept the externality effectively outside the park, despite the fact that the FPS took over its administration.

From wilderness dispute to the administrative duty of the Sámi Parliament – judicial framing

The doctrine of protection of old phases of forest succession was a source of great frustration to the FPS. The means of logging in forests reserved for economic use were to be scaled down, while demand for timber increased due to increased capacity in the plants in southern Lapland. At the beginning of the 1980s, the FPS was not willing to increase the area of conserved forests. The first major conflict over forestry in old forests resulted in establishment of Wilderness areas, where both reindeer herding and, to the great satisfaction to the FPS, also moderate forms of forestry were allowed.[46]

In forestry disputes from the 1980s onwards, the Sámi statements varied in their ecological depth; sometimes it was 'nature in its original state' that was to be protected,[47] but most statements, and all the formal institutions, the Sámi Delegation[48] and the Reindeer Herders' Association, stressed the need to protect reindeer herding. The Sámi voices still supporting forestry were marginalised. The fundamental difference had survived; in southern Finland, environmental values were the decisive driver for conservation, in Lapland it was the interests of competing means of living. The question had a cultural side to it, as herding was seen as a carrier of Sámi culture. Both aspects, the economic and the cultural, caused Sámi to protest about protection of predators in the conservation areas.[49] For one part, the Sámi looked to nature conservation for a preserved and saved resource zone; the questions of nature conservation were issues of compensation and securing usage rights.[50]

Some information exists on Sámi perception of wilderness areas. Initially, the chairperson of the Sámi delegation, Pekka Aikio, employed a judicial framing, protesting that the land rights issue remained outside the scope of the committee planning the wilderness areas.[51] The two wilderness areas in the Skolt

46. Parpola and Åberg 2009, pp. 354–56, 371–74, 382–86.
47. Kitti 1980a; Kitti 1980b.
48. The precedessor of the Sámi parliament, est. 1972, the Sámi Delegation was an elected organ for the Sámi, with an advisory mandate.
49. Nyyssönen 2000, pp. 186–90.
50. Nyyssönen 2000, p. 250.
51. Erämaakomitean mietintö 1988:39, passim.

Sámi administrative area seem to have been taken as a positive phenomenon, as areas for traditional usage forms. One reason for positivity seems to have been that, even though allowed, forestry had not been practised in the wilderness areas for many decades, while the Skolt Sámi are allowed to utilise their privately owned forests commercially. Wilderness legislation and the extreme geographical location of the forests thus provided assets to the Skolt Sámi in the competition for natural resources.[52]

One aspect beginning to de-stabilise the managemental peace was the different views of one central actor, reindeer. She no longer stood automatically among the species to be protected, but as the one who simultaneously benefits from and threatens nature protection, by becoming 'too many'. Segments of forest science sustained the fixed idea of reindeer as a harmful animal to the forest and pasture ecosystem. This was an unquestioned truism in early forestry research, articulated throughout the twentieth century, and recently increasingly under the biodiversity paradigm, faced with increasing evidence of the weakening pasture ecosystem.[53] As foresters utilised the idea of multiple use, representing forests as a resource for forestry and herding, the herder community, researchers favourable to their cause and the media have continued to seek faults in forestry (fragmentation of pasture areas, weakening of winter pastures etc.). Conflict has polarised and locked since the parties employ numerous frames (economic and cultural sustainability of Sámi communities vs. ecological sustainability of the Sámi means of living; health of the ecosystem vs. cultural rights and self-determination of the IP). Mainstream research stresses the impact of both to the ongoing change in the pasture ecosystem and externalities, among them industrial land-use forms and increased grazing pressure.

As the Sámi Parliament began to give statements on protection of nature in the Sámi Homeland, the sources of knowledge changed: the Sámi used studies on law, which meant that the issue was framed as a question of rights; that conservation plans must not violate herding and pasturing rights; that the legal foundation of the administration and rules of the natural reserves must be judicially solid in relation to the rights of the Sámi; and that it must be considered whether the case opened a possibility to air claims and/or point to violation of cultural rights.[54] The Sámi were to be reserved a right to build

52. Itkonen 2017, pp. 89–92.

53. Nyyssönen 2022, passim.

54. The archive of the Sámi Parliament of Finland, Statement 391/D.a. 9, 8.9.2005: Metsäntutkimuslaitoksen Kolarin tutkimusaseman lausuntopyyntö ratkaisuvaihtoehdoista, jotka koskevat suojelun ja alueen poronhoidon yhteensovittamista Mallan luon-

constructions and houses, according to the needs of traditional means of living and according to the conventions of Sámi culture, in the conservation areas in the Sámi Homeland.[55] The level of protection of the right to reindeer herding, fishing and hunting is of constant interest.[56] The role and the connected administrative grids of the Sámi Parliament as administrator of cultural autonomy guide the framing. The clear particularistic agenda in the ecopolitics of the Sámi Parliament has not gone unnoticed in prior research.[57]

The judicial frame marginalised ecological studies, forestry science and biology.[58] The predator question had also reached a crisis level, something that the Sámi Parliament has acted on: a right to remove individual predators was to be given to the Sámi as well.[59] In the most recent years, the conception of PAs as protective zones for reindeer (herding) has become a truism; part of this perception is the fixed position of forestry in the new constellation, i.e. culprit that diminishes the biomass of arboreal lichen and therefore consists a risk for the winter grazing of reindeer. Herder testimony of lesser need for artificial feeding in the PAs and research results showing bigger biomass of arboreal lichen in the conserved forest areas back up the conception.[60]

Conclusions

The frames concerning conservation have changed over the research period: the Sámi/herder benefit frame was joined, though not overtaken by, a more science-driven and more challenging indigenous rights frame. Common to

---

nonpuistossa. https://dokumentit.solinum.fi/samediggi/?f=Dokumenttipankki%2FAloitteet%2C%20esitykset%2C%20lausunnot%20ja%20muut (accessed 23 Dec. 2020.

55. The archive of the Sámi Parliament of Finland, Statement 508/D.a.4, 2.10.2009, Saamelaiskäräjien lausunto luonnonsuojelulain muuttamisesta. https://dokumentit.solinum.fi/samediggi/?f=Dokumenttipankki%2FAloitteet% 2C%20esitykset%2C%20lausunnot%20ja%20muut (accessed 23 Dec. 2020).

56. The archive of the Sámi Parliament of Finland, Statement 566/D.a.3.2007, 4.10.2007, Lausunto Lemmenjoen kansallispuiston hoito- ja käyttösuunnitelmasta, https://dokumentit.solinum.fi/samediggi/?f=Dokumenttipankki%2FAloitteet%2C%20esitykset%2C%20lausunnot%20ja%20muut (accessed 23. Dec. 2020)

57. Compare Berglund 2006, pp. 97–98.

58. Berglund 2006, pp. 103–08.

59. The archive of the Sámi Parliament of Finland, Statement 508/D.a.4, 2.10.2009, Saamelaiskäräjien lausunto luonnonsuojelulain muuttamisesta. https://dokumentit.solinum.fi/samediggi/?f=Dokumenttipankki%2FAloitteet% 2C%20esitykset%2C%20lausunnot%20ja%20muut (accessed 23 Dec. 2020).

60. Yle.fi/uutiset/3-11257874 (accessed 1 April 2020)

these frames was high insistence on usage rights. In the last phase of inquiry, the Sámi voices turned more challenging, not towards conservation, but towards competing land-use forms and to the sufficiency of Sámi rights. Peace is still prevalent within the protected areas, as forestry remains excluded from the PAs, and the PAs have turned into vehicles for monitoring the protection of other rights as well.

The way the committees framed the park planning processes with reference to the conservation practices of multiple use was one of the guarantees of success in the planning; the economic frames were shared, there was no dissonance in this way of approaching the PAs. The cause of environment, ecology and nature enjoyed a different weight for different actors. It was seldom referred to by the herders, but it did not amount to a hindering factor either, as herding was secured through the shared economically-framed approach and the environmental frame included herding in the entities to be protected. More decisively, all three frames, herding, environmental and that of Sámi survival, shared a negative view about efficient forestry. Differing interests and identity frames therefore matched sufficiently.[61]

The sources reflect the views of the group of Sámi herders and politicians with access to the PA planning and the framings of the state officials, who were behind the gathering of Sámi opinions and drafting the focus of the reports. The sources reveal an economic gaze, one wishing to see reindeer as a viable part of the landscape and income structures. Such instrumentality, and the success in restricting forestry outside the PAs, nuances the most victimised positions ascribed to IP in studies of PAs. Those with lesser standing received a voice in an issue that had long been handled without hearing them. At times, the institutional setting and a number of frames favoured their cause and the Sámi positioned successfully in relation to the competing frames.[62] The matters of the sufficiency of the protection of northern nature, of Sámi rights and their inclusion in PA management can be and are further discussed in research.[63]

Each usage of frame is linked to other frames and to larger cultural beliefs.[64] That the Sámi relation to their environments appears in the sources to be mostly economic, based on conservation of ecological services and natural resources, the productive capacity of nature,[65] is in part due to the mentioned

61. Kennedy, Brown and Butler 2021, passim.

62. Creed, Langstraat and Scully 2002, p. 45.

63. Heinämäki, Herrmann and Neumann 2014, passim.

64. Creed, Langstraat and Scully 2002, p. 45.

65. Ahonen 1997.

grid of sources. The grid has led to under-communicating arguments about cultural rights, of which there are only glimpses in the 1970s. The whole story frame stressing economy and fate of the means of living was coupled with identity frames of cultural survival and ethnicity of the Sámi. These identity frames enabled the Sámi actors to combine the individual, usually seen (in the context of Lapland) as anti-protectionist,[66] and the collective values and interests, neither of which negated conservation.

As conservation areas are established and the conserved everyday (use) of the areas begins, usually peace ensues. The most important reason is that the Finnish conservation legislation is straightforward on reindeer herding: it is allowed in national parks and wilderness areas. The level of protection of Sámi usage rights from the state and park administration and legislation was long considered sufficient.[67] The Finnish model resembles dispossessive examples from settler colonies only superficially; for example the oft-heard criticism about protecting unproductive peripheries, used and settled by IP,[68] do in practice apply in Finnish case. In addition, the state could be criticised of minimal effort at the inclusion of local voices and of Sámi in the planning and administration of the PAs. But, Sámi presence or livelihoods in the PAs were only regulated, not denied, and the encroachment on subsistence forms was meagre.

The 'failure' of PAs to qualify fully as colonial in the Finnish context does not imply that state policies concerning the Sámi could not be deemed colonial – there are recent examples of encroaching on Sámi fishing rights in the river Deatnu/Teno in the name of protecting the Atlantic salmon[69] and Sámi conservation policies, turning sour to the protection policies dictated from the 'south', are articulated under conditions of perception of diminishing, fragmented and threatened areas for traditional means of living and under a sharpened tone demanding self-determination. The criticism reveals another aspect of the conservation history of Finland: conservation has not stopped industrial development or resource extraction outside the PAs. This affects the Sámi as well and these matters are experienced as colonial by them. Prospecting and gold-digging in Lemmenjoki national park, forestry, hydropower projects, the tourism industry and reindeer herding have all left marks, to a varying degree, on nature in Upper-Lapland.

66. Kennedy, Brown and Butler 2021, p. 613.

67. https://yle.fi/uutiset/3-11695935 (accessed 11 Jan 2021).

68. Adams 2005, p. 129.

69. Toivanen and Cambou 2021, p. 58.

Bibliography

Adams, Michael. 2005. 'Beyond Yellowstone? Conservation and indigenous rights in Australia and Sweden'. In G. Cant, A. Goodall and J. Inns (eds). *Discourses and Silences, Indigenous Peoples, Risks and Resistance*. Christchurch: Department of Geography, University of Canterbury. pp. 127–38.

Ahonen, S. 1997. *Mitä on suojeltu, kun on suojeltu luontoa? - Käsitehistoriallinen tarkastelu suomalaisesta luonnonsuojelusta välillä 1880–1983*. Master's thesis, University of Helsinki, Faculty of Agriculture and Forestry, Department of Limnology and Environmental Protection. http://urn.fi/URN:NBN:fi-fe20011409

Äikäs, T. 2019. 'Religion of the past or living heritage? Dissemination of knowledge on Sámi religion in museums in Northern Finland'. *Nordisk Museologi* 3: 152–68.

Allard, Christina. 2017. 'Nordic legislation on protected areas: How does it affect Sámi customary rights?' In L. Elenius, C. Allard and C. Sandström (eds). *Indigenous Rights in Modern Landscapes, Nordic Conservation Regimes in Global Contexts*. London and New York: Routledge. pp. 9–25.

Andersson, R-H. 2019. 'Re-indigenizing National Parks: Toward a theoretical model of re-indigenization'. *Dutkansearvvi dieđalaš áigečála* 3: 65–83.

Andersson, Rani-Henrik, Boyd Cothran and Saara Kekki. 2021. 'Traditional Indigenous Knowledge and nature protection'. In R-H. Andersson, B. Cothran and S. Kekki (eds). *Bridging Cultural Concepts of Nature: Indigenous People and Protected Spaces of Nature*. Helsinki: Helsinki University Press. pp. 1–25. https://doi.org/10.33134/AHEAD-1-1.

Andreassen, Lars Magne. 2001. 'Makten til å kategorisere. Samepolitikk og vern av natur'. In A. Schanche (ed.) *Naturressurser og miljøverdier i Sámiske områder: forvaltnings- og forskningsutfordringer*, Dieđut 2. Kautokeino: Sámi instituhtta. pp. 134–53.

Andreassen, Lars Magne. 2002. 'Ka galt har vi gjort når det må vernes? Dilemmaer ved etablering av nasjonalpark i lulesamisk område'. In S. Andersen (ed.) *Sámisk landskap og Agenda 21, Kultur, næring, miljøvern og demokrati*, Dieđut 1. Kautokeino: Sámi instituhtta. pp. 115–28.

Berglund, Eeva. 2006. 'Ecopolitics through ethnography: The cultures of Finland's forest-nature'. In A. Biersack and J. B. Greenberg (eds). *Reimagining Political Ecology*. Durham & London: Duke University Press. pp. 97–120.

Creed, W.E.D., J.A. Langstraat and M.A. Scully. 2002. 'A picture of the frame: Frame analysis as technique and as politics'. *Organizational Research Methods* 5: 34–55.

Elenius, Lars. 2017. 'Introduction'. In L. Elenius, C. Allard and C. Sandström (eds). *Indigenous Rights in Modern Landscapes, Nordic Conservation Regimes in Global Contexts*. London and New York: Routledge. pp. 1–6.

Erämaakomitean mietintö. 1988: 39.

Getzner, M., M.L. Vik, E. Brendehaug and B. Lane. 2014. 'Governance and management strategies in national parks: Implications for sustainable regional development'. *International Journal of Sustainable Society* 6 (1–2): 82–101. https://doi.org/10.1504/IJSSOC.2014.057891

Green, Carina. 2009. *Managing Laponia, A World Heritage as Arena for Sámi Ethno-Politics in Sweden*. Ph.D. diss. Faculty of Arts, Uppsala University.

Hallikainen, V., M. Jokinen, M. Parviainen, L. Pernu, J. Puoskari, S. Rovanperä and J. Seppä. 2006. 'Inarilaisten käsityksiä metsätaloudesta ja muusta luonnonkäytöstä'. *Metsätieteen Aikakauskirja* 4: 453–74.

Hallikainen, V., T. Helle, M. Hyppönen, A. Ikonen, M. Jokinen, A. Naskali, S. Tuulentie and M. Varmola. 2008. 'Luonnon käyttöön perustuvat elinkeinot ja niiden väliset suhteet Ylä-Lapissa'. *Metsätieteen aikakauskirja* 3: 191–219.

Heinämäki, L., T.M. Herrmann and A. Neumann. 2014. 'Protection of the culturally and spiritually important landscapes of Arctic Indigenous Peoples under the Convention on Biological Diversity and first experiences from the application of the Akwe:Kon Guidelines in Finland'. *Yearbook of Polar Law* 6: 189–225.

Itkonen, Panu. 2017. 'Land rights as the prerequisite for Sámi culture, Skolt Sámi's changing relation to nature in Finland'. In L. Elenius, C. Allard and C. Sandström (eds). *Indigenous Rights in Modern Landscapes, Nordic Conservation Regimes in Global Contexts*. London and New York: Routledge. pp. 83–95.

Joutsamo, Esko. 2008a. 'Kansallis- ja luonnonpuistojen synty'. In H. Telkänranta (ed.) *Laulujoutsenen perintö, Suomalaisen ympäristöliikkeen taival.* Helsinki: Suomen luonnonsuojeluliitto. pp. 80–89.

Joutsamo, Esko. 2008b. 'Taistelu Metsä-Lapista'. In H. Telkänranta (ed.) *Laulujoutsenen perintö, Suomalaisen ympäristöliikkeen taival.* Helsinki: Suomen luonnonsuojeluliitto. pp. 128–33.

Kansallispuistokomitean mietintö 1976: 88.

Kennedy, B.P.A., W.Y. Brown and J.A. Butler. 2021. 'Frame analysis: An inclusive stakeholder analysis tool for companion animal management in remote Aboriginal communities'. *Animals* 11: https://doi.org/10.3390/ani11030613.

Kitti, J. 1980a. 'Luonddusuodjalanguovllut viidanit'. *Sápmelaš* 9–11/1980.

Kitti, J. 1980b. 'Láddelaš luonddodukti ávžžutus: "Maŋimuš sámeguovlu Anaris lea šaddamin heđiid sisa"'. *Sápmelaš* 9–11/1980.

Koilliskairatoimikunta, Ehdotukset Koilliskairan suojelualueesta, Koitilaisen luonnonpuistosta ja Oulangan kansallispuiston laajentamisesta. Toimikunnan mietintö 1972, I osa.

Koilliskairatoimikunta, Porotaloustyöryhmä, (Alaruikka, Aikio, Alakorva, Savukoski, Hokajärvi). 1972. 'Selvitys poronhoidon asemasta Koilliskairan, Koitilaskairan ja Oulangan alueilla sekä poronhoidon jatkuvuuden turvaaminen näiden mahdollisten kansallis- ja luonnonpuistojen tai muun tyyppisillä rauhoitusalueilla'. *Poromies* 5/1972.

Luonnon- ja kansallispuistokomitean mietintö 1953: 9.

Markkula, I., M.T. Turunen and S. Kantola. 2019. 'Traditional and local knowledge in land use planning: insights into the use of the Akwé: Kon Guidelines in Eanodat, Finnish Sápmi'. *Ecology and Society* 24: https://doi.org/10.5751/ES-10735-240120.

Niemi, Seija A. 2018. *A Pioneer of Nordic Conservation: The Environmental Literacy of A.E. Nordenskiöld (1832–1901).* Ph.D. diss. Faculty of Humanities, University of Turku.

Nordin, Åsa. 2008. 'Livsformteorier och renskötsel – En ny utgångspunkt'. In P. Sköld (ed.) *Människor i Norr, samisk forskning på nya vägar.* Umeå: Vaartoe – Centrum för samisk forskning, Umeå Universitet. pp. 175–94.

Nyyssönen, Jukka. 2000. 'Murtunut luja yhteisrintama, Inarin hoitoalue, saamelaiset ja metsäluonnon valloitus 1945–1982'. Licentiate thesis. Faculty of Humanities, University of Jyväskylä.

Nyyssönen, J. 2022. 'A competing harvester, stakeholder, and environmental threat: Positioning reindeer in Metla forestry research'. *Society & Animals*: https://doi.org/10.1163/15685306-bja10076

Parpola, Antti and Veijo Åberg. 2009. *Metsävaltio, Metsähallitus ja Suomi 1859–2009*. Helsinki: Edita.

Raitio, Kaisa. 2008. *'You Can't Please Everyone' – Conflict Management Practices, Frames and Institutions in Finnish State Forests*. Ph.D. diss. Faculty of Social Sciences and Regional Studies, University of Joensuu.

Reimerson, Elsa. 2021. 'Discourses of decentralization: Local participation and Sámi space for agency in Norwegian protected area management'. In R-H. Andersson, B. Cothran and S. Kekki (eds.) *Bridging Cultural Concepts of Nature. Indigenous People and Protected Spaces of Nature*. Helsinki: Helsinki University Press. pp. 61–93: https://doi.org/10.33134/AHEAD-1-3.

Sara, Inker-Anni. 2019. *Whose Voice? Understanding Stakeholder Involvement in Law Drafting affecting Sámi Reindeer Herding*. Ph.D. diss. Faculty of Humanities and Social Sciences, University of Jyväskylä.

Schanche, Audhild. 2001. 'Innledning'. In A. Schanche (ed.) *Naturressurser og miljøverdier i Sámiske områder: forvaltnings- og forskningsutfordringer*, Dieđut 2. Kautokeino: Sámi Instituhtta. pp. 3–19.

Suominen, Teuvo. 2008. 'Urpo Häyrinen, suojeluhehtaarien ykkönen?' In H. Telkänranta (ed.) *Laulujoutsenen perintö, Suomalaisen ympäristöliikkeen taival*. Helsinki: Suomen luonnonsuojeluliitto. p. 85.

Telkänranta, Helena (ed.) 2008. *Laulujoutsenen perintö, Suomalaisen ympäristöliikkeen taival*. Helsinki: Suomen luonnonsuojeluliitto.

Toivanen, Reetta and Dorothée Cambou. 2021. 'Human rights'. In C.P. Krieg and R. Toivanen (eds.) *Situating Sustainability: A Handbook of Contexts and Concepts*. Helsinki: Helsinki University Press., pp. 51–62: https://doi.org/10.33134/HUP-14-4.

Vahtola, Jouko. 1999. 'Petsamo Suomen tieteen tutkimuskohteena'. In J. Vahtola and S. Onnela (eds.) *Turjanmeren maa, Petsamon historia 1920–1944*. Rovaniemi: Petsamo-Seura. pp. 485–507.

Veistola, Tapio. 2008. 'Luonnonsuojelu EU-Suomessa'. In H. Telkänranta (ed.) *Laulujoutsenen perintö, Suomalaisen ympäristöliikkeen taival*. Helsinki: Suomen luonnonsuojeluliitto. pp. 238–43.

Vincent, Eve and Timothy Neale (eds). 2016. *Unstable Relations: Indigenous People and Environmentalism in Contemporary Australia*. Crawley: UWAP Scholarly.

Internet

Statements, The archive of the Sámi Parliament of Finland. Statement 391/D.a. 9, 8.9.2005: Metsäntutkimuslaitoksen Kolarin tutkimusaseman lausuntopyyntö ratkaisuvaihtoehdoista, jotka koskevat suojelun ja alueen poronhoidon yhteensovittamista Mallan luonnonpuistossa: https://dokumentit.solinum.fi/samediggi/?f=Dokumenttipankki%2FAloitteet%2C%20esitykset%2C%20lausunnot%20ja%20muut (accessed 23 Dec. 2020).

Statement 566/D.a.3.2007, Lausunto Lemmenjoen kansallispuiston hoito- ja käyttösuunnitelmasta 4 Oct. 2007: https://dokumentit.solinum.fi/samediggi/?f=Dokumenttipankki%2FAloitteet%2C%20esitykset%2C%20lausunnot%20ja%20muut (accessed 23 Dec. 2020).

Statement 508/D.a.4, 2.10.2009, Saamelaiskäräjien lausunto luonnonsuojelulain muuttamisesta: https://dokumentit.solinum.fi/samediggi/?f=Dokumenttipankki%2FAloitteet%2C%20esitykset%2C%20lausunnot%20ja%20muut (accessed 23 Dec. 2020).

Yle.fi/uutiset/3-11257874 (accessed 1 April 2020).

Samediggi.fi (accessed 25 Oct. 2022).

Chapter 8

WOLVES AND THE FINNISH WILDERNESS: CHANGING FORESTS AND THE PROPER PLACE FOR WOLVES IN TWENTIETH-CENTURY FINLAND

Heta Lähdesmäki

Introduction

These days, if wolves roam close to human settlements, Finnish people often argue that there is something unnatural in their behaviour. This is the case especially in western Finland, where wolf packs are now being observed after a long period of absence.[1] Not all local people have welcomed wolves as their neighbours. For instance, in a demonstration held in 2018 in Vaasa, a coastal city in Ostrobothnia in western Finland, one participant stated that 'wolves do belong to Finland but in the wilderness. There is no wilderness in Ostrobothnia.'[2] Wolves have been connected to the wilderness in many countries and regions in the world.[3] In some areas, the notion that the wolf belongs to the wilderness is old: for instance, historian Aleksander Pluskowski has argued that there was a persistent conceptual link between wolves and the wilderness in Britain and

1. Nowadays, wolves exist almost everywhere in Finland, outside the reindeer herding area in the north. Now, more wolves live in western Finland than in the eastern part of the country, which was previously their main habitat. *Suomen susikannan painopiste yhä enemmän lännessä*, Natural Resources Institute Finland.

2. 'Susimielenosoitus keräsi torin täyteen väkeä Vaasassa – "Veikkaan, että äänemme kuultiin tänään"', *Yle* 11 June 2018.

3. See for instance, Lopez 1978, pp. 140–44; Mech 1995, p. 271; Tonnaer 2020.

doi: 10.3197/63824846758018.ch08

Scandinavia during the Middle Ages.[4] In this article, I look into the idea that wolves belong to the wilderness and trace its history in the Finnish context.

By looking at newspaper reports, magazine articles and contemporary literature, I ask when, how and why the wilderness came to signify the proper place for wolves in Finland. I argue that it is a relatively new and controversial notion connected to various social and environmental changes.

Wolves that move about near humans

The Finnish wolf population declined at the turn of the twentieth century, but a few hundred years ago, there could have been almost 1,400 wolves in Finland.[5] According to historian Jouko Teperi, wolves inhabited the whole country before the decline. During the nineteenth century, the human population grew, human settlements spread, logging increased and cultivated land expanded. Consequently, peoples' and wolves' living spaces began to partly overlap. Teperi writes how Finns often observed wolves near human settlements, especially during winter. Nineteenth-century newspapers wrote about peoples' encounters with wolves on roads and how wolves, their tracks and kills were seen in cattle pastures and near peoples' houses. Finns did not appreciate this kind of behaviour but still described it as typical for wolves.[6]

Similar attitudes were voiced during the first part of the twentieth century. Even though it is likely that only a few wolves lived in Finland at this time, the zoological book *Suomen luurankoiset (Vertebrata Fennica)* published in 1909 claimed that, during summer, wolves inhabit 'deserted forests', but during winter, particularly in freezing weather, often approached human-inhabited areas in large packs.[7] The newspaper *Wasabladet* reported in June 1901 that wolves had headed out to human-inhabited areas in the Oulu region in northern Ostrobothnia early that year in search of better food.[8] In these descriptions, wolves did not live permanently near humans, but visiting built-up areas was presented as an established habit. Newspapers continued to write about such visits as the twentieth century progressed.[9]

---

4. Pluskowski has studied the period from the eighth to the mid-fourteenth century. Pluskowsk 2006.
5. Aspi et al. 2006, pp. 1569, 1572; Jansson et al. 2014, p. 2.
6. Teperi 1977; Lähdesmäki & Ratamäki 2015.
7. Salovaara 1930, p. 111; Kivirikko 1940, p. 24. All translations are made by the writer.
8. Wargar, *Wasabladet* 18 July 1901, p. 3.
9. Lähdesmäki 2020b.

Wolves were depicted near humans also in illustrations published in hunting magazines. In 1908, *Suomen metsästyslehti*, published by the Finnish Hunting Association, contained an illustration with three wolves approaching a house (see Figure 1).[10] In 1912 and 1916, the magazine's successor, *Metsästys ja Kalastus*, published pictures with wolves walking near human settlements (Figures 2 and 3).[11] All these pictures depict wintertime. We do not know what the magazine's publishers wished to express to readers with these illustrations. They may have caused worry and fear but might have also strengthened the idea that it is common for wolves to visit built-up areas.

Figure 1. Three wolves approach a closed yard. Source: Suomen metsästyslehti 10/1908, p. 301.

According to local newspapers and hunting magazines, wolves were quickly chased away and killed, if possible, whenever they were observed near built-up areas.[12] This was possible because of the wolf's legal status before 1973 as a harmful animal that could be killed anywhere by anyone.[13] What is significant is that, even though people wanted to kill them, wolves' behaviour was not described as unnatural or abnormal in the first half of the twentieth

10. *Suomen metsästyslehti* 10/1908, p. 301.

11. *Metsästys ja Kalastus* 2/1912, p. 41; *Metsästys ja Kalastus* 2/1916, p. 3.

12. Lähdesmäki 2020b.

13. See *Hunting Act*, 290/1962.

Figure 2. Two wolves walk past barns. Source: Metsästys ja Kalastus 2/1916, p. 3.

Figure 3. A lone wolf looks at a group of houses. Source: Metsästys ja Kalastus 2/1912, p. 41. This illustration is a copy of the painting 'Samotny Wilk' (The lone wolf) by Polish artist Alfred Wierusz-Kowalski (1849–1915).

century. Visits were sometimes condemned as irritating and unfortunate but, most of the time, newspapers and magazines merely stated the fact in a declarative way. Sometimes, the articles pondered why wolves had roamed to built-up areas. Cold weather and hunger were presented as reasons, and sometimes the behaviour was stated to be 'normal' and 'natural' for wolves.[14]

During the second half of the twentieth century, the way people understood wolves' behaviour changed. It was no longer 'normal' for wolves to visit or live part-time in areas inhabited by humans. At the beginning of the century, wolves that moved about near humans were occasionally called *brave*.[15] Later similar behaviour was described as *bold* and *arrogant*. For instance, in April 1949, newspaper *Lapin Kansa* stated that a wolf individual had been 'arrogant' because it had circled a village in Lapland.[16]

I argue that these arrogant wolves were considered to be stepping out of place by visiting human settlements. Animal geographers use the concept of being 'out of place' to highlight, as Chris Philo writes, how 'animals often squeeze out of the places – or out of the roles that they are supposed to play in certain places – which have been allotted to them by human beings'.[17] Wolves were arrogant and out of place because the way Finns defined the cultural environment had changed to exclude their presence. According to a new sociospatial order, wolves and humans did not coexist.

Wolves out-of-placeness was connected to the changing ways in which Finns perceived uncontrollability and fear. Wolves' presence caused people to be afraid of attacks on domestic animals and even humans.[18] As Finnish society modernised, fear of predators was no longer something that was part of normal human life. People were more able to control their lives than, for instance, in the nineteenth century when, according to Teperi, Finns perceived wolves as an uncontrollable force of nature.[19] As uncontrollability and fear became 'abnormal' factors in human life, wolves' proper place was discursively marginalised and pushed away from built-up areas to the wilderness.

---

14. See for instance 'Susia Vuoksenniskalla', *Metsästys ja Kalastus* 3/1912, p. 29; 'Toivo Kuparinen: Petoja rajaseudun maisemassa', *Metsästys ja Kalastus* 9/1960, p. 337; Siivonen 1956, p. 144.

15. See Mela and Kivirikko 1909, p. 26; 'Sudet', *Suomen metsästyslehti* 2/1908, p. 56.

16. Hukat röyhkeinä Sodankylässä, *Lapin Kansa* 20.4.1949, p. 1.

17. Philo 1995, p. 656.

18. About the fear of wolves attacking and killing humans in Finland, see for instance Lähdesmäki 2020a.

19. Teperi 1977, pp.73–75; Lähdesmäki 2020b, p. 208.

'In the wilderness, one hears the howl of the wolf'

Wolves and wilderness had already been connected in the nineteenth century. For instance, in 1880, the newspaper *Uusi Suometar* described how wolves make their dens in the dense forests of the wilderness.[20] Still, I argue that this connection was not emphasised until the twentieth century. The popular zoological book *Suuri nisäkäskirja* (1956), associated wolves clearly with wilderness while describing their shyness:

> This quality [shyness] is present only when a wolf meets a human; I doubt if a wolf is afraid of anything but humans in the wilderness ... You can hear its call and try to catch it in vain as if you were catching the wind. However, this is only in the wilderness, where it has a secure livelihood and is not teased by hunger.[21]

Later the book stated that '[i]t is a true call of the wilderness, warning and gloomy when wolves howl'.[22] Forty-one years later, the popular zoological book *Suomen luonto* (1997) also connected the howl with wilderness: 'The howl of wolves is a moving, unforgettable song of the wilderness.'[23]

Wolves were connected to the wilderness also through illustrations in popular zoological books. Wolves, their traces and dens were now primarily depicted in wilderness-like environments such as forests or wetlands that had no apparent signs of human activity.[24]

In 1960, the hunting magazine *Metsästys ja Kalastus* argued clearly that wolves and other large predators belong to the forests and that their proper place was not near humans.[25] Similarly, an opinion piece writer stated in the newspaper *Helsingin Sanomat* in 1997 that large predators belong to Finnish national parks and 'other uninhabited areas'.[26] It seems that many contemporary Finns shared this view: The vast majority of participants of two opinion surveys conducted in the late 1990s considered the wolf to be a resident of wilderness areas. Some hoped wolves would live somewhere with enough wilderness for them to live without causing too much harm to humans, such as

---

20. Sudet Pomarkussa, *Uusi Suometar* 7.6.1880, p. 6.
21. Siivonen 1956, pp. 145–46.
22. Ibid., p. 147.
23. Kojola 1997, p. 157.
24. See Siivonen 1956, p. 139; Siivonen 1972, pp. 101, 145.
25. Toivo Kuparinen, 'Petoja rajaseudun maisemassa', *Metsästys ja Kalastus* 9/1960, p. 340.
26. 'Laki ja suojelu sekaisin', *Helsingin Sanomat* 26 July 1997, p. A17 (opinion piece, written by Ulla Hyttinen, Vihti).

nature reserves and national parks in Lapland and Eastern Finland.[27] In reality, Finnish conservation areas did not guarantee protection until the latter part of the century: wolf hunting became forbidden in all Finnish nature reserves in the 1990s, but is still allowed in some national parks and Lapland due to reindeer husbandry in the area.[28]

Even though people occasionally saw wolves near built-up areas and even in peoples' yards at the end of the twentieth century, many Finns believed that wolves actually *need* wilderness areas to thrive.[29] For instance, the popular zoological book *Suomen eläimet I* (1983) stated that wolves need peace and protection from humans.[30] Finns were not alone with this view: biologist David Mech argues that the idea that wolves require habitat free of human influences to survive is a misconception and that equating wolves with wilderness results from wolves being exterminated in most areas except wilderness.[31]

The idea that wolves should only live in the wilderness and are out of place elsewhere was linked to geographical othering, described by Chris Philo and Chris Wilbert. In Western countries, population centres such as cities are suitable areas for pets or companion animals, while rural areas are for livestock animals. This leaves unoccupied lands beyond the margins of human settlement and agriculture, often called the wilderness, for wild animals such as wolves. In reality, the world does not divide into clear and separate areas for different species to live in.[32] Wolf individuals visiting human settlements challenged the notion that wolves live in the wilderness. Still, this notion was not entirely imaginary. Because the number of wolves living in Finland was small, fewer wolves were seen near humans during the twentieth century compared to the previous century. The visits were infrequent, making this behaviour seem abnormal. Wolves might have also changed their behaviour: According to present-day wolf researchers, when wolves are persecuted, they tend to become more wary of humans. Also, wolves that are cautious toward humans might live longer

27. Lumiaro 1997, pp. 12, 15, 41; Vikström 2000, pp. 77–78.

28. *Nature Conservation Act* (1096/1996), 13 §.

29. See for instance, Ritva Liikkanen, 'Salametsästäjät ilmeisesti estävät Suomen susikantaa kasvamasta', *Helsingin Sanomat* 15 Jan. 1998, p. A5; Lumiaro 1997, pp. 12, 15, 41. About the visits, see Ahti Takala, 'Susi vieraili Lappajärvellä', *Metsästys ja Kalastus* 4/1982, p. 57; 'Punahilkkaoireyhtymä vaivaa Kuhmossa', *Suomen Luonto* 3/1997, pp. 45–46.

30. Pulliainen 1983, p. 193.

31. Mech 1995, p. 271; Fritts et al. 2003, p. 300.

32. Philo and Wilbert 2007, pp. 10–11.

than less cautious wolves.[33] Therefore, it might have been beneficial for wolves to live further away from humans or at least stay unnoticed.

Wilderness imagined

Interestingly, as the idea that wolf is a wilderness species grew stronger in Finland, areas where wolves could live outside human influence grew fewer. Population growth and expansion of human habitation altered the landscape shared by these two species. The number of people living in Finland grew from over 2.6 million at the beginning of the twentieth century to over 5.1 million in 2000.[34] At the same time, the agricultural land area increased. The field area was at its maximum in 1968, at over 2.6 million hectares. [35] Also, the expansion of the road and rail networks fragmented the landscape. The road network expanded from 32,000 km in 1915 to 77,993 km in 2000.[36] The railway often ran parallel to the roads, narrowing ecological corridors. In many countries, roads have also raised the risk of wolves being killed by people.[37] In Finland too, people have been able to use roads when killing wolves.[38]

Finns have used and altered forests for hundreds of years, for instance, for the production of tar and for slash-and-burn cultivation. During twentieth century, the use of forests intensified and mechanised. In the 1950s and 1960s, logging was mechanised, forest roads were built for forest machines, forests were fertilised, brushwood was poisoned and forest plantations were planted. Forest growth was increased by draining swamps and wetlands. Intensive forestry has changed and is changing Finnish forests in many ways: the age structure of forests has become younger, vast wilderness areas have vanished and forests are fragmented. At the same time, forest biodiversity has decreased.[39] Even though Finland has of the largest forest areas in the European Union (almost eighty per cent of land area is forests), Finnish forests are mainly commercial.[40]

33. Fritts et al 2003, pp. 300–02.
34. Statistics Finland, population structure.
35. Salokangas 2003, p. 673; Vihola 2004, pp. 332–39.
36. Salokangas 2003, p. 676; Kiiskinen 1999, pp. 9–10, 13, 100–01; Finnish Road Statistics 2009; The Finnish Railway Statistics 2016.
37. Fritts et al 2003, p. 301; Laaksonen 2013, p. 46.
38. Lähdesmäki 2020b, p. 174.
39. Myllyntaus 1999, pp. 101–02; Puttonen 2006, p. 79; Leikola 2006, pp. 87–88.
40. Nègre 2022; *Forest Resources in Finland.*

After these societal and environmental changes, Finland had no true wilderness where wolves could roam outside human influence. Similar changes also happened elsewhere: wolves have lost their habitat throughout Europe and North America due to human activity and, nowadays, most of the world's wolves live near humans.[41] Still, urbanisation – which took place relatively late in Finland compared to many other western countries – helped to create the idea that there could still be vast wilderness areas in the country. Finnish society was agrarian until the 1950s but only two decades after that, in 1970, more than half the Finnish population lived in cities and towns.[42]

The meaning of the Finnish word for wilderness, *erämaa*, has also changed. The idea of wilderness is connected to the Romantic period, nature conservation movement and nationalism. Previously feared, untamed nature began to be seen as aesthetically beautiful in the western world at the end of the eighteenth century. During the first half of the twentieth century, the English term *wilderness* started to signify vast, pristine areas and authentic, pure nature, often without a past or human presence. The idea that wilderness has no human past is, as several environmental historians have argued, problematic. Many areas perceived as wilderness have been inhabited for millennia by humans and human influence can be seen almost everywhere in the world.[43]

However, the notion of *erämaa* has included humans. The word's etymology is connected to the hunting and fishing culture practised from the prehistorical era to at least the Middle Ages in the area today called Finland. Historically, *erämaa* has meant an area where a house or village community has the right to hunt. Still, as geographer Jarkko Saarinen notes, different people in Finland have had and still have conflicting notions about wilderness. According to Saarinen, there are at least three discourses on wilderness in present-day Finland: *conserved wilderness* functions as a 'biodiversity container', *traditional wilderness* is an economic resource and subject of usage especially to local people, and *touristic wilderness* is a commercialised site for touristic activities. The idealised images of wilderness are associated with northern and eastern Finland.[44] When wolves are associated with wilderness, it is seen as *conserved*

41. Breitenmoser 1998, pp. 279–85; Fritts et al. 2003, p. 300.

42. Haapala 2007, pp. 50–59.

43. See Nash 1982 (1967); Coates 1998, pp. 10, 157, 177; Coleman 2004, pp. 135; Tonnaer 2020. Bill McKibben has written about the effects of peoples' actions on the rest of the nature in his book *The End of Nature* (1989). Recently, researchers in human sciences have used the concept of Anthropocene to describe human influence on Earth and its ecosystems. See Kress and Stine 2017.

44. Saarinen 2019.

wilderness, a notion that draws on the Western and Anglo-American views on wilderness and wild nature and is seen as a place without humans.

'Wolves do not need wilderness to thrive'

Even though the idea that wolves need wilderness still exists in Finland, it has been challenged by many Finns, at least since the late twentieth century. In 1972, a lone wolf living in the Tavastia region in southern Finland got plenty of media attention since wolves had not been observed in this area for a long time. Local nature conservation organisations defended the wolf's right to live there. In an opinion piece published in the newspaper *Hämeen Sanomat*, the organisations' representatives stated that 'although wolf is a predator, it is not a predator that could not live together with humans in the same landscape'.[45] *Suomen luonto* (1997) told its readers that wolves could permanently live near humans, not just visit human settlements occasionally, and even have dens in the cultural landscape.[46] Present-day wolf researchers state that wolves do not depend on wilderness areas; if allowed, they could thrive in areas densely populated by humans.[47] The cultural landscape is multispecies: as geographers Jennifer Wolch, Kathleen West and Thomas Graines remind us, even cities are not merely human, and many animals that are perceived as wildlife can live permanently or part-time in cities.[48]

Why was this behaviour, moving around human settlements and visiting urban areas, described again as 'normal' at the end of the twentieth century? I argue that how Finns viewed the wolf as a species played a vital role. During the twentieth century, cultural notions of wolves underwent significant changes. Environmental historian John McNeill has written about how wolves graduated from varmints to noble savages from 1960 to 1990 in the Western world.[49] This change also took place in Finland. Wolf conservation in Finland began in 1973, outside the reindeer herding area, which covers about 36 per cent of

45. 'Puheenvuoro: Hämeen suden puolesta', *Hämeen Sanomat* (written by Pentti Andsten from Kanta-Häme nature conservation association and Tuomo Raunistola from Luonto-Piiri) 30 Oct. 1972, p. 5.
46. Kojola 1997, pp. 157–58.
47. Mech 1995; Fritts et al 2003; Boitani 2003.
48. Wolch, West and Gaines 1995, pp. 736–37, 753. In many non-Western cities, wildlife is not seen as out-of-place.
49. McNeill 2001, p. 340. See also Mech 1995, p. 271.

the country's surface.[50] After this, wolves were seen as creatures with rights – for instance, the right to exist in Finland.[51] According to an opinion survey conducted in the late 1990s, some Finns in favour of wolf protection thought wolves should be protected even if living near humans.[52]

'Problem wolves' and the right to kill wolf individuals

I argue that the need to highlight the wolf's reputation as a wilderness species during the second part of the twentieth century was linked to justification of their killing. Before wolf conservation began in Finland in the 1970s, the wolf was classified as a harmful animal, and legislation encouraged people to hunt wolves down. The idea that the wolf is a wilderness species reinforced the already existing right to kill wolves upon meeting them, so there was no need to emphasise it. After wolves became a protected species, humans could no longer kill them as freely as before.[53] In the 1990s, when Finland became a member of the European Union, wolf conservation became stricter outside the reindeer herding area and more bureaucratic.[54] In order to kill a wolf legally, people needed to get special licence to make an exception to the protection. Also, one needed to justify the need to kill individual wolves that were under protection as a species.

Even though the species was protected, many Finns did not wish to share their living environment with wolves. In 1984, Matti Valtonen, the executive director of the Hunters' Central Organisation, stated in the magazine *Metsästys ja Kalastus* what many Finns thought – that no one wants a wolf in their backyard.[55] From the late twentieth century onward, newspapers often called wolves who visited peoples' yards *häirikkösusi*. The word is roughly translated as a troublemaker or nuisance wolf and is related to the English term

---

50. *Decree No. 749/1973 on Wolf Protection.*
51. Lähdesmäki 2020b.
52. Lumiaro 1997, p. 15.
53. The decree on wolf protection did not actually restrict the killing of wolves that much outside the reindeer herding area. It was possible to kill wolves legally during certain months in some of the Eastern municipalities, where most of the wolves existed during the 1970s and 1980s. It was also possible, under certain circumstances, to kill wolves elsewhere in Finland; for instance, if wolves caused damage or their numbers became too numerous. *Decree No. 749/1973 on Wolf Protection.*
54. *Council Directive 92/43/EEC on the conservation of natural habitats and of wild fauna and flora.*
55. 'Suurpetomme lisääntyvät', *Metsästys ja Kalastus* 1/1984, p. 31

problem animal.[56] The concept of *häirikkösusi* was introduced in the 1980s. In 1982, Viljo Sivonen wrote in the hunting magazine *Metsästys ja Kalastus* about wolves that killed dogs while visiting peoples' yards. Sivonen was annoyed at this behaviour but also because the state did not compensate for the kills. In his article, he addressed the officials, lawmakers, game researchers and wolf conservationists: '[I]f you cannot help us through the law, then come and remove these troublemaker wolves from our midst next autumn. We are running out of means!'[57] In the 1990s, this term was used in abundance, for instance, in the country's biggest newspaper *Helsingin Sanomat*.[58]

I argue that the concept of *häirikkösusi* was connected to the notion that wolves belong to the wilderness. It was, in a way, another way of saying that wolves that move about near humans are not normal. It was also a way to morally justify the human urge to keep wolves away from built-up areas, even by killing them. In the 1990s, some municipalities in eastern Finland applied for killing permits for wolves that had moved around human settlements.[59] Many Finns considered the killing of these so-called troublemaker wolves justifiable. Even some wolf conservationists and environmental organisations supported the killing of problem wolves.[60]

Conclusions

In this article, I have traced the history of the notion that wolves belong to the wilderness and shown how it is a relatively new and controversial idea in Finland. As the twentieth century progressed, the wolf became a symbol of the wilderness to many Finns, and their presence near humans changed from being criticised yet 'typical' and 'normal' to being seen as 'bold' and 'arrogant'

---

56. On problem animals, see Linnell et al. 1999, p. 698

57. Viljo Sivonen, 'Ilomantsin susisyksy 1981', *Metsästys ja Kalastus* 5/1982, p. 62.

58. In the newspaper *Helsingin Sanomat*, the term and its synonyms were used in six news reports written by journalists and in five opinion pieces. See, for instance, Ritva Liikkanen, 'Metsästyssäännöksiä ei aiota muuttaa', *Helsingin Sanomat* 22 Aug. 1997, p. A8; 'Susi ei ole häirikkö syntyessään', *Helsingin Sanomat* 13 Oct. 1997, p. A9 (opinion piece written by Timo Helle, chair of The Finnish Association for Nature Conservation, and Tuomas Rantanen, secretary-general of The Finnish Nature League).

59. See, for instance, 'Talon nurkissa kiertelevälle sudelle tappotuomio Nurmeksessa', *Helsingin Sanomat* 30 Jan. 1999, p. A5.

60. 'Paluu 1800-luvun petopolitiikkaan', *Helsingin Sanomat* 27 Jan. 1997, p. A9 (opinion piece written by Riku Lumiaro, Vantaa); 'Mika Parkkonen', 'Suojelijat eivät vastusta pihaan tulevan suden tappoa', *Helsingin Sanomat* 26 April 1997, p. A17.

behaviour. Cultural notions and their historical roots are important to examine because they have affected the lives of both people and wolves. If people have perceived wolves to be out of place, it has historically resulted in wolves being chased or killed.

Wolves were associated with wilderness partly because Finns did not want to share their living spaces with wolves. There were many reasons for that, mainly related to the fact that, as historian Jon T. Coleman puts it, wolves and humans have competed for space and calories. When moving about near humans, wolves have had access to domestic animals. Coleman writes about wolves' history in America, where European-Americans understood livestock as their property.[61] Finns have felt similarly. Also, the possibility that wolves could harm humans has made them unwanted and feared visitors in the cultural landscape.[62]

Interestingly, the connection between wolves and the wilderness was also made because some Finns believed that wolves need wilderness areas to thrive, that it was essential for them to live far away from humans and population centres. As I wrote earlier, according to present-day wolf researchers, this notion is a misconception. Still, wolves are more threatened in open terrain and areas highly developed and used by humans. Also, the absence of humans in the wolves' habitat is an essential factor for wolves to thrive.[63] For wolves, wilderness-like areas would therefore be advantageous, but paradoxically, at the same time as the idea that the wolf is a wilderness species strengthened, the Finnish environment underwent changes that meant that the areas that could be called wilderness became fewer. As I write earlier, environmental historians have brought light to the fact that people have shaped the shared landscape so that there are practically no places left for wolves to be completely isolated from human influences in Finland and elsewhere.

Wolf conservation that began in the 1970s has made it possible for wolves to proliferate, form packs and spread out into old territories in Finland. As I related at the beginning of this chapter, not all Finns have welcomed them. It has been difficult for many to co-exist with wolves. Compared to many other European countries, the Finnish wolf population is relatively small: the estimate number of wolves living in Finland was under 300 individuals in March 2022.[64] Strict conservation has worsened disputes between interest

---

61. Coleman 2004, pp. 35–36.
62. See for instance, Lähdesmäki 2020a; Lähdesmäki 2020b.
63. Mech 1995; Breitenmoser 1998, pp. 281–82; Barber-Meyer et al. 2021.
64. *Suomen susikannan painopiste yhä enemmän lännessä.*

groups and politicised wolves' presence in the Finnish landscape.[65] In previous decades, wolves have returned to many European countries after shorter or longer periods of absence. Therefore, Finns are not alone in debating whether wolves belong to the cultural landscape or to the imagined or real wilderness. For instance, in Finland's neighbouring countries Norway and Sweden, where recovering wolf populations are nowadays protected, many people believe that wolves do not belong to cultural landscape or that their numbers should be kept minimal there.[66] In Russia, where the species is not protected, the general attitude toward wolves is more negative, and many Russians view the wolf as a competitor.[67] Wolves' presence creates heated debates in Portugal, Spain, Switzerland and Germany too. As Michaela Fenske and Bernhard Tschofen point out, '[w]olves are a keystone species that exemplify humanity's relation to what is called nature and their return generates powerful debates about what nature actually is and how much it is needed or should be permitted to exist'.[68] In order to live in a multispecies world, we need to try to understand wolf and other wildlife debates. To do so, we need to know the histories of the ways we have perceived animal places.

---

65. Lähdesmäki 2020b; Ratamäki 2009.

66. Sjölander-Lindqvist 2009; Figari and Skogen 2011.

67. Kirilyuk 2020.

68. Fenske and Tschofen 2020, p. i. On the return of wolves in many European countries, Michaela Fenske and Bernhard Tschofen (eds), *Managing the Return of the Wild. Human Encounters with Wolves in Europe.*

Bibliography

Aspi, Jouni, Eeva Roininen, Minna Ruokonen, Ilpo Kojola and Carles Vilà. 2006. 'Genetic diversity, population structure, effective population size, and demographic history of the Finnish wolf population'. *Molecular Ecology* 15: 1561–76.

Barber-Meyer, S.M., T.J. Wheeldon and L.D. Mech. 2021. 'The importance of wilderness to wolf (*Canis lupus*) survival and cause-specific mortality over 50 years'. *Biological Conservation* **258** (109145): 1–13.

Boitani, Luigi. 2003. 'Wolf conservation and recovery'. In David Mech and Luigi Boitani (eds). *Wolves. Behavior, Ecology and Conservation.* Chicago: University of Chicago Press. pp. 317–344.

Breitenmoser, Urs. 1998. 'Large predators in the Alps: The fall and rise of man's competitors'. *Biological Conservation* **83** (3): 279–89.

Coates, Peter. 1998. *Nature. Western Attitudes since Ancient Times.* Cambridge: Polity Press.

Coleman, Jon T. 2004. *Vicious: Wolves and Men in America.* New Haven: Yale University Press.

Council Directive 92/43/EEC on the conservation of natural habitats and of wild fauna and flora.

Decree No. 749/1973 on Wolf Protection.

Fenske, Michaela and Bernard Tschofen (eds). 2020. *Managing the Return of the Wild: Human Encounters with Wolves in Europe.* London: Routledge.

Figari, Helene and Ketil Skogen. 2011. 'Social representations of the wolf'. *Acta Sociologica* 54 (4): 317–32.

Finnish Railway Statistics 2016. Statistics from the Finnish Transport Agency 9/2017. Finnish Transport Agency. https://www.traficom.fi/sites/default/files/media/file/rautatietilasto_2016.pdf

Finnish Road Statistics 2009. Statistics from the Finnish Transport Agency, 2/2010. Transport and Tourism. https://www.doria.fi/bitstream/handle/10024/121692/lti_2010-02_978-952-255-009-5.pdf?sequence=1&isAllowed=y.

Forest Resources in Finland. Ministry of Agriculture and Forestry of Finland. https://mmm.fi/en/forests/forestry/forest-resources (accessed 30 Sept. 2022).

Fritts, Steven H., Robert O. Stephenson, Robert D. Hayes and Luigi Boitani. 2003. 'Wolves and humans'. In David Mech and Luigi Boitani (eds). *Wolves. Behavior, Ecology and Conservation.* Chicago: University of Chicago Press. pp. 289–316.

Gade-Jørgensen, Inge and Rie Stagegaard. 2000. 'Diet composition of wolves Canis lupus in east-central Finland'. *Acta Theriologica* **45** (4): 537–47.

Haapala, Pertti. 2007. 'Kun kaikki alkoi liikkua'. In Kai Häggman et al. (eds). *Suomalaisen arjen historia. 3. Modernin Suomen synty.* Porvoo: Weilin+Göös. pp. 47–63.

Hunting Act, 290/1962

Jansson, Eeva, Jenni Harmoinen, Minna Ruokonen and Jouni Aspi. 2014. 'Living on the edge: Reconstructing the genetic history of the Finnish wolf population'. *BMC Evolutionary Biology,* 14: 1–20.

Kaartinen, Salla, Miska Luoto and Ilpo Kojola. 2010. 'Selection of den sites by wolves in boreal forests in Finland'. *Journal of Zoology* 281: 99–104.

Kiiskinen, Hanne. 1999. *Kurvit Kohdallaan: Tielaitos 200 Vuotta.* ed. by Kimmo Levä. Kangasala: Mobilia.

Kivirikko, K.E. 1940. *Suomen selkärankaiset. Vertebrata Fennica.* Porvoo: WSOY.

Kojola, Ilpo. 1997. 'Susi'. In Petri Nummi et al. (eds) *Suomen luonto. Eläimet. Nisäkkäät.* Porvoo: Weilin + Göös. pp. 154–58.

Kress, W. John and Jeffrey K. Stine (eds). 2017. *Living in the Anthropocene: Earth in the Age of Humans.* Washington, D.C.: Smithsonian Institution.

Laaksonen, Mervi. 2013. *Susi.* Helsinki: Maahenki.

Leikola, Matti. 2006. 'Metsien hoidon ja käytön kehittyminen 1900-luvulla Suomessa'. In Riina Jalonen, Ilkka Hanski, Timo Kuuluvainen, Eero Nikinmaa, Paavo Pelkonen, Pasi Puttonen, Kaisa Raitio and Olli Ilari Tahvonen (eds). *Uusi metsäkirja.* Helsinki: Gaudeamus. pp. 84–90.

Linnell, John D.C., John Odden, Martin E. Smith, Ronny Aanes and Jon E. Swenson. 1999. 'Large carnivores that kill livestock: Do "problem individuals" really exist?' *Wildlife Society Bulletin* **27** (3): 698–705.

Lopez, Barry. 1978. *Of Wolves and Men.* New York: Charles Scribner's Sons.

Lumiaro, Riku. 1997. *Onko sudella olemassaolon mahdollisuutta Suomessa – ihmisten suhtautuminen suteen.* Master's thesis. University of Helsinki.

Lähdesmäki, Heta. 2020a. 'The memory of a shared past. From human-wolf conflicts to coexistence?' In Marlis Heyer and Susanne Hose (eds). *Encounters with Wolves: Dynamics and Futures.* Serbski institut Budyšin Bautzen. pp. 33–48.

Lähdesmäki, Heta. 2020b. *Susien paikat. Ihminen ja susi 1900-luvun Suomessa.* Doctoral thesis. Nykykulttuurin tutkimuskeskuksen julkaisuja 127. Jyväskylän yliopisto: Jyväskylä.

Lähdesmäki, Heta and Outi Ratamäki. 2015. 'Kykenemmekö luopumaan susifetissistä? – Kriittinen luenta suomalaisesta susihistoriasta'. In Juha Hiedanpää and Outi Ratamäki (eds). *Suden kanssa.* Rovaniemi: Lapin yliopistokustannus. pp. 16–41.

McNeill, John R. 2001. *Something New Under the Sun. An Environmental History of the Twentieth Century World.* Lontoo: Penguin Books.

McKibben, Bill. 1989. *The End of Nature.* New York: Random House.

Mech, L. David. 1995. 'The challenge and opportunity of recovering wolf populations'. *Conservation Biology* 9: 270–78.

Myllyntaus, Timo. 1999. 'Aarniometsistä puupeltoihin: Metsät Suomen taloudessa'. In Timo Soikkanen (ed.), *Ympäristöhistorian näkökulmia. Piispan apajilta trooppiseen helvettiin.* Turku: University of Turku. pp. 88–103.

Nash, Roderick. 1982 [1967]. *Wilderness and the American Mind.* New Haven: Yale University Press.

Nature Conservation Act (1096/1996)

Nègre, Francois. 2020. The European Union and forests. https://www.europarl.europa.eu/ftu/pdf/en/FTU_3.2.11.pdf (accessed 30 Sept. 2022).

Philo, Chris. 1995. 'Animals, geography, and the city: Notes on inclusions and exclusions'. *Environment and Planning. D, Society & Space* **13** (6): 655–81.

Philo, Chris and Chris Wilbert. 2000. 'Introduction'. In Chris Philo and Chris Wilbert (eds). *Animal Spaces, Beastly Places.* London: Routledge. pp. 1–35.

Pluskowski, Aleksander. 2006. *Wolves and the Wilderness in the Middle Ages*. Woodbridge: Boydell.

Pulliainen, Erkki. 1983. 'Susi'. In Ilkka Koivisto et al. (eds.) *Suomen eläimet I*. Espoo: Weilin + Göös. pp. 188–95.

Puttonen, Pasi. 2006. 'Metsien käytön historia. Johdanto'. In Riina Jalonen, Ilkka Hanski, Timo Kuuluvainen, Eero Nikinmaa, Paavo Pelkonen, Pasi Puttonen, Kaisa Raitio and Olli Ilari Tahvonen (eds). *Uusi metsäkirja*. Helsinki: Gaudeamus. pp. 79–81.

Ratamäki, Outi. 2009. *Yhteiskunnallinen kestävyys ja hallinta suomalaisessa susipolitiikassa*. Doctoral thesis. Yhteiskuntatieteellisiä julkaisuja nro 94. Joensuu: University of Joensuu.

Saarinen, Jarkko. 2019. 'What are wilderness areas for? Tourism and political ecologies of wilderness uses and management in the Anthropocene'. *Journal of Sustainable Tourism* 27 (4):. 472–87.

Salokangas, Raimo. 2003. 'Itsenäinen tasavalta'. In Seppo Zetterberg et al. (eds), *Suomen historian pikkujättiläinen*. Porvoo: WSOY. pp. 597–689.

Salovaara, Hannes. 1930. *Eläinten maailma. Kuvia ja kuvauksia. I Imettäväiset*. Helsinki: Otava.

Siivonen, Lauri. 1956. *Suuri nisäkäskirja*. Helsinki: Otava.

Siivonen, Lauri (ed.) 1972. *Suomen nisäkkäät II*. Helsinki: Otava.

Sjölander-Lindqvist, Annelie. 2009. 'Social-natural landscape reorganised: Swedish forest-edge farmers and wolf recovery'. *Conservation and Society* 7 (2): 130–40.

Suomen susikannan painopiste yhä enemmän lännessä, Natural Resources Institute Finland, News 21 June 2022: https://www.luke.fi/fi/uutiset/suomen-susikannan-painopiste-yha-enemman-lannessa (accessed 26 Sept. 2022).

'Susimielenosoitus keräsi torin täyteen väkeä Vaasassa – "Veikkaan, että äänemme kuultiin tänään"'. *Yle*, video. 11 June 2018: https://yle.fi/uutiset/3-10249050 (accessed 1 Aug 2018).

Teperi, Jouko. 1977. *Sudet Suomen rintamaiden ihmisten uhkana 1800-luvulla*. Helsinki: Suomen historiallinen seura.

Tonnaer, Anke. 2020. 'The story of *Wanderwolf*. A contested tale on the re-emergence of 'new wilderness' in the Netherlands'. In Michaela Fenske and Bernhard Tschofen (eds). *Managing the Return of the Wild. Human Encounters with Wolves in Europe*. London: Routledge. pp. 47–61.

Vikström, Saara. 2000. *Suurpetoasenteet poronhoitoalueen eteläpuolisessa Suomessa vuonna 1999*. Masters' thesis. University of Oulu.

Vihola, Teppo. 2004. 'Pärjääkö pienviljelys?' In Ann-Carin Östman, Matti Peltonen, Teppo Vihola and Heikki Rantatupa (eds). *Suomen Maatalouden historia. 2, Kasvun ja kriisien aika 1870-luvulta 1950-luvulle*. Helsinki: SKS. pp. 198–203.

Statistics Finland, population structure. https://www.tilastokeskus.fi/tup/suoluk/suoluk_vaesto_en.htmll (accessed 27 April. 2023)

Whatmore, Sarah. 2002. *Hybrid Geographies. Natures, Cultures, Spaces*. London: Sage.

Wolch, Jennifer; Kathleen West and Thomas Gaines. 1995. 'Transspecies urban theory'. *Environment and Planning D: Society and Space* **13** (6): 735–60.

Chapter 9

ALL QUIET ON THE EASTERN FRONT? THE FINNISH ARMY AND WILDLIFE DURING WORLD WAR TWO

Mauri Soikkanen and Simo Laakkonen

World War Two was a macroscale war but its history is made of microhistories. In Finland, World War Two consisted of three different but connected wars. The Soviet Union attacked Finland in November 1939, starting the Winter War, which ended in March 1940 with a peace treaty signed in Moscow. During the Continuation War, 1941–1944, Finland together with Nazi Germany, attacked the Soviet Union to regain lost areas. Hostilities were ended by an armistice agreement that obliged Finland to oust German forces from the northern part of the country, which took place in the Lapland War of 1944–1945. In this chapter we will especially focus on the environmental history of the Continuation War, due to its length and relevance for our theme. After the rapid advancement of the Finnish army at the beginning of this war in summer 1941, the front line stagnated in East Karelia for almost three years. In summer 1944 the Red Army launched a major attack, which the Finnish Army was eventually able to stop; the Continuation War ended in September 1944 and Finland retained its independence, capitalist economy and democratic institutions.[1]

Despite the significant overall impact of wars on societies and nature, environmental studies have tended to focus on peacetime processes. Consequently, environmental history is a new approach to the study of the largest war that has taken place on Earth so far. This article addresses the mobilisation of natural resources in Finland during World War Two. More specifically, we study the following question: what role did hunting and fishing have in the war zone during the Continuation War? Depictions do exist in the memoirs of soldiers and officers from different countries, but hardly any historical studies

1. See Laakkonen and Vuorisalo (eds) 2007.

doi: 10.3197/63824846758018.ch09

have been conducted on these themes despite their importance.[2] This article is probably the first attempt to review the extent of hunting and fishing activity in a war zone during World War Two, and its significance both for military personnel and wildlife populations.[3]

In the interwar period the more-than-1,500-kilometre-long Finno-Soviet border extended from the Baltic Sea in the south to the Barents Sea in the north. This chapter addresses a region named East Karelia, situated west of the White Sea and Lake Onega and north of Lake Ladoga, Europe's largest freshwater lake. It is a sparsely inhabited wilderness of thousands of lakes, immense marshes and conifer forests. It is part of the taiga – that is, the boreal forest zone that extends from Siberia in the east to the Nordic Countries in the west. East Karelia covers over 100,000 square kilometres, an area larger than that of contemporary Portugal or Austria.

Our focus is on traditional game and fish species of Northern Europe. We attempt to discover how it was possible to hunt and fish in the conditions of a world war, and what significance this had for Finnish troops as well as for fish and game populations in the area. Our article focuses primarily on the everyday aspects of wartime woodcraft, including hunting, trapping and fishing practices and catches; as well as if attempts were made to regulate hunting and fishing during war so that game and fish stocks would not be destroyed. This article is based on archival material in addition to wartime professional publications and journals and reminiscences of Finnish veterans.[4] Interviews with almost 600 veterans who served in the 14th division of the Finnish army from 1941 to 1944 are collected in the archives of the Finnish Literature Society. We refer to these interviews anonymously.

We argue that, while the Finnish military administration endeavoured to protect game and fish resources at the outset of the war, wartime conditions and ruthless exploitation of wildlife meant there could only be modest success. The oasis of wildlife that the Finnish army found in East Karelia was soon lost.

---

2. The impact of WWII on marine fisheries and fish stocks has been studied. See, for example, Holm 2012.
3. None of the following three edited volumes on the environmental history of WWII addresses fishing or hunting on or behind the frontlines: Laakkonen, Tucker and Vuorisalo (eds) 2017; Laakkonen, McNeill, Tucker and Vuorisalo (eds) 2019; Robertson, Tucker, Breyfogle and Mansoor (eds) 2020.
4. See Soikkanen 1999.

Map 1. East Karelia occupied by the Finnish army in 1941–1944. The map shows the eastern Finnish border both before and after the war as well as the conquered territory at its largest. Marked on the map is Kentjärvi where the biological field station of the University of Helsinki was located during the war.

The Karelian Autonomous Soviet Socialist Republic

In territorial terms, since the late Middle Ages East Karelia had belonged first to Novgorod, then to Imperial Russia and finally to the Soviet Union, but the population of the area was historically Finno-Ugric, consisting of mainly of Karelians and Finns, who spoke rather similar languages. Russian-speaking Slavic populations began to migrate to the area in greater numbers after the eleventh century. The majority of verses in the Finnish national epic, *The Kalevala*, were collected from the Finnic inhabitants of East Karelia. At the turn of the nineteenth and twentieth centuries, East Karelia became the object of Karelianism, an idealistic nationalism and romanticism that captured the imagination of Finland's cultured elite. Karelianism gradually turned into a political ideal aimed at annexing East Karelia to Finland.[5] But, from 1917, when Finland achieved independence and the Soviet Union was founded, the border between the countries was gradually closed. Despite this, East Karelia remained a mythical region in the minds of many Finns, including those who spent time fishing and hunting in the wilderness.

The local populace of the East Karelian wilderness was skilled in hunting and fishing. Traditionally hunting and fishing provided them with a significant addition to their livelihood and income.[6] After the October Revolution, the Soviet state established the Karelian Autonomous Soviet Socialist Republic, or KASSR, in East Karelia. Communist power brought curtailment of liberties, confiscation of private property and collectivisation. In fishing *kolkhozy,* only professional fishermen had access to effective fishing equipment.[7] Hunting with modern weapons was allowed only by small body of professional hunters. After the Soviet state confiscated weapons, local inhabitants had to resort to hunting with traps and other primitive equipment.[8] All this hampered flourishing traditional hunting and fishing in East Karelia.

The KASSR attempted to regulate hunting by juridical means as well. A hunting statute laid out by the Soviet Republic in 1935 offered complete protection to moose and Finnish forest reindeer, which had survived in East Karelia only. Also completely protected were American imports – the muskrat, the silver fox and the American mink. Soviet hunting seasons were slightly more liberal than Finnish hunting seasons. Hunting bear, wolf and wolverine was

5. See Nygård 1978; Näre and Kirves (eds) 2014.
6. Nieminen 1998, p. 281; Hämynen 1998, pp. 175–76.
7. Halme 1943, p. 98; Siltamaa 1984, p. 50.
8. Airaksinen 1943, p. 42.

permitted year-round for professional hunters. The KASSR also established nature reserves to protect the Finnish forest reindeer and specific wilderness areas. Leningrad's academic institutions carried out zoological and ichthyological research in the region.[9]

Stalin's paranoia was notorious for being directed at minority nationalities. In the KASSR, Finnic people like Finns, Karelians, Vepsians and Ingrians were particular targets of Stalinist persecution.[10] In 1935, an order was given to NKVD to purge minority nationals from the Soviet border zone and in 1937–1938 the great terror was carried out. Due to mass deportations, sentences and killings, traditional hunting and fishing waned or stopped completely in Karelian villages located on the Soviet side of the border.[11]

The brutal developments on the Soviet side of the border were noticed by Finnish game researchers. They noted through snow-track counts conducted at the border that the reduction of hunting in East Karelia was rapidly reflected in an increase in the populations of game animals at the Finno-Soviet frontier.[12] To put it rather crudely, it can be argued that one of the few positive results of the Stalinist terror was the increase in game and perhaps also fish populations in some parts of the Soviet Union, and partially in some neighbouring countries as well.

Use and management of East Karelia

Stalin's policies at the end of the 1930s had a profound impact on the abundance of game and fish in East Karelia. When Finnish soldiers crossed the border into East Karelia in late June 1941, the region's waters were full of fish and its game stocks ample.[13] In the eyes of the Finnish soldiers, the game populations in the

---

9. 'Karjalan ASNT:n metsästyssäännöt ja metsästysajat'. Decree no. 247, 1935. (Photocopy).

10. During the terror of 1937–1938, about 11,000 Finnic people were sentenced, of whom about two thirds were of Finnish or Karelian nationality. Over 80% of those sentenced were executed. For information on the history of the KASSR, see Kangaspuro 2000.

11. Autio 2002, pp. 261–63.

12. 'Riistakannan vahvistuminen vv. 1937–38 karjalaisvainojen eräänä seurauksena', Uusi Suomi, 31 Jan. 1943.

13. Halme 1943, p. 100; Lampio 1946, p. 21; Lauri Siivonen, 'Itä-Karjalan riistatilanne ja metsästysolot. Kertomus Valtion tieteelliselle Itä-Karjalan toimikunnalle', Suomen kirjallisuuden seura, 30 Oct. 1942. (Photocopy); Interview by Simo Laakkonen with Sakari Laakkonen (b. 1919), veteran and retired building contractor, 1 Mar. 2006, Porvoo, Finland.

area appeared fabulously plentiful, especially as several of the species that existed in annexed Karelia had been hunted to near or utter extinction in Finland. The Finnish forest reindeer had been hunted to extinction in Finland by the early 1800s, and the moose almost suffered the same fate a hundred years later. Of fur-bearing animals, the Eurasian beaver was extinct and wolverine, lynx, wolf, bear and pine marten were encountered in Finland only in frontier lands. Large birds of prey had also almost disappeared, thanks to bounty hunting and general enmity towards predators.[14] Overfishing, building of dams, massive log floating and urban-industrial pollution had annihilated stocks of valuable fish species such as salmon, trout and white fish in Finnish Karelia.[15] In brief, democratic and capitalist Finland had taken particularly poor care of its game and fish stocks prior to World War Two.

Finland needed the natural resources of the annexed territory and so, by the time Finnish troops took East Karelia, plans for the region's use and management were ready. The East Karelian military administration undertook this management. The plan was that ownership and use of the annexed territory's lands and waters would be more tightly restricted than in Finland. The sizes of new plots would only be large enough to ensure subsistence farming. Forests would be doled out in parcels large enough to meet the demands of households, but not large enough to bring in additional income through commercial use. Ample debate had taken place in Finland over the private ownership of waters, and it was decided that the best solution in East Karelia was that the state would retain ownership.[16] East Karelia was to become a new societal laboratory for Greater Finland, in which public administration would guarantee rational management of natural resources.

On this foundation, the military administration of East Karelia also planned the hunting and fishing in the area and their regulation. Already by August 1941 the first general directive for both the civilian populace and soldiers was given. It also regulated fishing and hunting:

The populace of Karelia must demonstrate commitment and initiative and begin gathering food stores for the upcoming autumn. Fishing is hereby declared free to all in all lakes and rivers; however, explosives and poisonous substances or drugging agents may not be used. At least one third of a fishing channel's breadth must always be left open. The hunting of other game, such as

14. See Lehikoinen 2007.

15. See Laakkonen and Bolotova 2021.

16. Laine 1982, pp. 282–83.

Figure 1. A young Finnish soldier and baby owls. At the front, soldiers took care of all kinds of pets, which shows the young men's desire to take care of living things even in dire conditions. Source: Photo Archives of the Finnish Defence Forces, photographer Tk-Keskimäki.

moose and wild boar, is also free, but the use of firearms without the approval of authorities is forbidden.[17]

There was an urgent need to tap the natural resources of East Karelia, as there was a continuous lack of foodstuffs on the front and the homefront. In particular, the winter of 1941–1942 was extremely difficult and lack of food also afflicted East Karelia. If at all available, fish and game offered a badly needed dietary supplementation.

Putting the natural resources of East Karelia to immediate use proved possible. At the time, Finland was an agrarian nation; the majority of Finnish soldiers were thus used to fishing or hunting to meet household needs. During wartime, everyone had weapons that could be used for hunting as well, and simple gear was often sufficient to catch fish. As early as the winter of 1941–1942, hunting and fishing detachments made up of skilled hunters and fishers were beginning to be established among the troops, and it was their task to bring in game for the army's use. The military administration gave more specific orders regarding hunting and fishing, patterned after the laws and regulations laid down in Finland. A decree made in the spring of 1943 established rules for all hunting aside from moose hunting. For instance, soldiers were required to obtain a personal written authorisation to hunt, a hunting licence.[18] The hunting seasons were in line with those established in Finnish homeland.

Moose – the prime game animal

Once the fronts had stabilised by the fall of 1942, widespread hunting began in East Karelia. The moose was the most desired catch, but most Finnish soldiers were unfamiliar with how to hunt it, since the moose had been hunted almost to extinction in the early 1900s, and the species had been completely protected up to 1933. Even since then, the take had been low: before World War Two, the annual take was only 500–600 moose a year in the entire country.[19] In East Karelia, however, moose were prolific.

According to the new edict issued by the Finnish military administration in autumn 1942, moose hunting was a licensed activity and allowed during

---

17. Proclamation No. 2 of the East Karelian military administration commander to the populace of East Karelia. Itä-Karjalan sotilashallinnon säädöskokoelma, 7–10, 1941. The Finnish military archives (Sota-arkisto, SA) are now part of the Finnish National Archives (henceforth SA).

18. Edict on hunting, Itä-Karjalan sotilashallinnon säädöskokoelma no. 30. 1943, SA.

19. Kairikko 1991, pp. 403–05.

Figure 2. During the Second World War, the majority of the Finnish military consisted of farmer boys skilled in the ways of the forest. This moose, shot in East Karelia, was quickly flayed and cut into pieces as well. These skills proved useful, especially during unlicensed moose hunts. Source: Photo Archives of the Finnish Defence Forces, photographer Tk-Keskimäki.

the period between 16 October and 16 November.[20] Moose-hunting licences were granted by the district chief from the military administration. He also established the licence areas after having agreed on the matter with the commander responsible for the military organisation. The military organisation then distributed the licences to entire units and to individual hunters within them. For instance, for the hunting season of fall 1943, units were granted permission to take 1,500 moose, and individual hunters within the units, 600 moose.[21] Thus, within the area managed by the East Karelia military administration, the

20. Edict on moose hunting, 4 Oct. 1942.

21. Edict on hunting in the area of the East Karelia military administration 30/43. Made 25 Aug. 1943 by Chief of Military Administration Olli Paloheimo. Itä-Karjalan sotilashallinnon esikunnan arkisto [Archives of the staff of the East Karelia military administration], SA.

size of the legal moose take that year was 2,100 animals – about four times as many as moose legally taken in Finland on an annual basis.

This is the official story, but the reality is that moose were taken in much greater numbers.[22] They were hunted year-round, as the soldiers found it difficult to understand that you could only shoot moose from October to November, yet you could shoot people year-round. Superiors also recognised this contradiction, and so in practice not much attention was paid to the protection orders given by the military administration. Moose taken outside the hunting season were normally noted on reports as having been killed 'in the minefields'.[23] Thus East Karelia became a true oasis of big game for licensed hunters and poachers as well.

As moose stocks diminished, air force assistance was sometimes called on to bring in supplementary food. Especially during wintertime patrolling expeditions, fighter pilots could readily spot herds of moose and deer in the white and open frozen swamplands and lakes, and it was easy to fell the necessary amount of game with fighter plane machine guns. The pilot would report the location of the take to the air base, and a scouting party would go to retrieve it. This method of hunting was also prohibited under threat of court martial, but monitoring this prohibition in the vast stretches of wilderness was difficult, if not impossible.[24]

Hunting and protection of Finnish forest reindeer

The Finnish forest reindeer is distinct from the domesticated reindeer in its larger size and preference for dense forest. Regrettably it had been hunted to extinction in Finland in the early nineteenth century, but did not receive protected status in Finland until 1913. When reports of this rare creature began to arrive from East Karelia at the beginning of the Continuation War, some enlightened officers, hunters and conservationists alike began immediately campaigning on behalf of the species in the war zone. As a result of the political pressure, the military administration of East Karelia fully protected the Finnish forest reindeer as early as October 1941,[25] which can be considered respectably rapid conservation policy-making in wartime conditions.

22. Siivonen, 'Itä-Karjalan riistatilanne ja metsästysolot. Kertomus Valtion tieteelliselle Itä-Karjalan toimikunnalle', Suomen kirjallisuuden seura, 30 Oct. 1942, p. 9.

23. See, Henttonen and Lappalainen 1991.

24. Interview with Väinö Pokela on March 2, 1999 in Helsinki; Interviews with veterans of the 14th division; Soikkeli, 'Hirvenmetsästystä hävittäjäkoneella Vienan Karjalassa'.

25. Ylänne 1941, pp. 257–58; 'Alkusanat', Suomen Luonto 2/1942.

Yet hunger drove the soldiers to it, and Finnish forest reindeer were hunted rather ruthlessly despite their protected status, especially in the winter of 1941–1942. They were shot later as well, even though the military administration stressed the protection of the species and threatened shooters of wild reindeer with court martials. Consequently, in the later reports of the hunting detachments, the Finnish forest reindeer no longer appeared; it was normally recorded as a 'small moose'.[26]

Fur-bearing animals

The populations of large fur-bearing animals such as wolves, wolverine, lynx and brown bear were much more sizable in East Karelia than in Finland.[27] However, the most common and most significant fur-bearing animal was a somewhat smaller species – the squirrel.[28] Yet, the squirrel was not only a fur-bearing animal; during the 1942 winter of famine, the soldiers cooked skinned squirrels in their mess kits and went so far as to pronounce them good, 'they just tasted a little pitchy'.[29]

Red fox was common in East Karelia, and it was also an important and valuable fur-bearing animal. Foxes were shot in fields as they stalked mice. Other furbearers were also hunted. Otter and pine marten always got hunters out the door. The skins of tundra hares were also set aside. The muskrat had spread to East Karelia by this point, and was in certain locales very abundant.[30] The European polecat and the European mink, rare in Finland at that time, were also among the animals hunted by soldiers.[31]

The abundant stock of fur-bearing animals induced hunters to rely on a variety of hunting methods. Hunting with poison was familiar to the elderly Karelians who had remained in the occupied territory. These old-timers instructed the Finnish soldiers in the art of poison trapping, as few of the latter had practised this in their homeland. As a result, during wintertime the walls

26. Soikkanen 1999, p. 28.

27. Lampio 1946, p. 26; Siivonen, 'Itä-Karjalan riistatilanne ja metsästysolot. Kertomus Valtion tieteelliselle Itä-Karjalan toimikunnalle', Suomen kirjallisuuden seura, 30 Oct. 1942, p. 7; Itä-Karjalan sotilashallinnon esikunnan arkisto, SA.

28. 'Yli 2 milj. oravaa ammuttu – 120 milj. ansaittu', Metsästys ja Kalastus 2 (1943): 62; Soikkanen 1999, p. 74.

29. Interviews with veterans from the 14th division.

30. Siivonen, 'Itä-Karjalan riistatilanne ja metsästysolot. Kertomus Valtion tieteelliselle Itä-Karjalan toimikunnalle', Suomen kirjallisuuden seura, 30 Oct. 1942, p. 18.

31. Hunting regulations of the KASSR, (no place or publisher) 1935.

Figure 3. The commander of the 14. division and an expert in wilderness warfare, the 'wilderness general' Erkki Raappana giving his Karelian Bear Dog a scratch after a successful bear hunt. The Second World War proved attritive to Finland's dog population. The Seskar Seal Dog, an ancient breed living in the outer archipelago of the Gulf of Finland went extinct. The populations of the Finnish Hound, Finnish Spitz and Lapponian Herder saw a massive decline. The Karelian Bear Dog population only survived due to studs brought in from East Karelia. Source: Photo Archives of the Finnish Defence Forces, photographer Tk-Rossi.

of the dugouts were strewn with drying fox skins, which were then sold to East Karelian wholesale traders if there was a dire need for money.[32]

However, the soldiers trusted more in the domestic market, since there was a general lack of all goods during wartime, meaning pelts fetched good prices. In the fur auction of 1943, an average of 1,500 Finnish mk was paid for a red fox pelt. Muskrat was a true luxury fur; the price rose to 300–400 mk for a single skin. In civilian life, you could earn more over a couple of weeks of muskrat hunting than you could breaking your back in the fields all summer. The skins of large predators were exceptionally desirable and valuable. A good bear skin brought in 5,000–7,000 mk.[33] Furthermore, the Finnish government

32. Kulha 2004.

33. 'Turkisten hinnat', *Metsästys ja Kalastus* 10 (1943): 285.

supported bounty hunting of large predators in the occupied territory, just as at home.[34] The hunting bounty for a wolf in East Karelia was 3,000 mk, 2,000 mk for a wolverine and 1,000 mk for a bear. Only half the sum was paid for pups, kits and cubs.[35]

The overall income from hunting small and big game could in individual cases be astounding, when one takes into consideration the earnings of soldiers at the time. In the summer of 1943, the per diem for soldiers at the front was 40 mk for a lieutenant, 30 mk for a sergeant first class and 16 mk for a private. The family at home received a so-called monthly war stipend, 850 mk for a wife and two children. Thus one good skin of a large predator could support a whole family for several months. The fur-bearing animals of East Karelia shielded the families of many Finnish soldiers from poverty during the war.

Profitable fishing

Fishing offered soldiers an easier-to-access dietary supplement than hunting, and was also a more popular and more profitable activity among the soldiers than hunting. Hooks and a bob were stowed in the backpacks of many soldiers and, when they returned from leave, nets and long lines were brought from home. Boats were sometimes found on the shores of the lakes; in other instances, log rafts were built to get out onto the water. If no 'normal' fishing equipment was at hand, soldiers made do with what was available – explosives.

The majority of explosives were distributed during the onslaught of autumn 1941, when there was no time to use normal fishing tackle. Hand grenades, mines, dynamite or TNT were used to build a piled-up charge that was given the moniker 'pioneer fishing rod'. A satchel charge with six kilos of explosives, built around a German-type stick hand grenade, was often used in fishing. This is how it was used:

> One man was at the upper oarlock, two men at one pair of oars and the best thrower of the group in the rear bench. When the rowers figured they had achieved full speed, the fellow in the back gave it all he got and threw a piled-up charge behind them. It took a moment before the charge reached the bottom and the fuse burned. Then Boom! The first time, it felt like the boat would burst too. But we didn't sink ourselves after all. Fish began rising to the surface

34. Bounty hunting was widely practised in Europe in past centuries. See, Pohja-Mykrä, Vuorisalo and Mykrä, 2005.

35. Itä-Karjalan sotilashallinnon esikunnan arkisto, SA.

> in a radius of about 100 yards. We got pike, whitefish, bream and big perch. Vendace was too small to collect.[36]

Explosives were commonly used in fishing during WWII, even though it had been prohibited in the general directive given in 1941. There was an attempt to root out this destructive form of fishing with education and monitoring as well, but it seems to have been in use until the summer of 1944, when the fishing waters of East Karelia were forced to be ceded.

The fishing administration

Great hopes were pinned on fishing in occupied East Karelia. After all, part of magnificent Lake Ladoga and hundreds of other large, fish-rich lakes were now at Finland's disposal. A collection was organised in the Finnish homeland to bring fishing gear to the occupied territory, and it produced nets, fykes and even a few seines. The highest authority on fishing management was the military administration's agency of fish trade. Buying and brokering goods from the fishermen was the exclusive right of the wholesale company established by the military administration, which also managed the trade in fur and meat.[37]

Fishing brought a significant supplement to the local economy. The local populace had a right to fish for their own needs in waters near their homes without a permit. In practice, this applied only to the Karelian populace, since the Russian populace of East Karelia had been interred in prison camps. One needed a permit for boating, and fishing was a regulated activity in those lakes where the enemy held the opposite shore. On some of the larger lakes, boating was prohibited because of the danger of partisans. The prohibition was an attempt to offer the populace security, but also to prevent any possible partisan contact.

Professional fishermen had come from the Finnish homeland to the occupied territory in hopes of better catches and greater earnings. However, the price for the fish was regulated.[38] In the summer of 1944, a total of 505 fishermen were working in 91 professional fishing groups under the military administration. Their fishing gear included about 100 seines, more than 2,000 various fykes and more than 4,000 nets. Professional fishing per se was organised in such a way that the military administration rented gear to the fishermen and retained a third of the catch as payment. Fishermen were also allowed to buy the nets provided by the military administration, and towards the end

36. Sopanen 1992, pp. 74–75.

37. Simonen 1971, p. 89.

38. *Paateneen Viesti*, 5 June 1942.

Figure 4. Lacking modern fishing equipment, soldiers utilised handy Stone Age techniques for fishing in the wilderness. This picture depicts the soldiers checking a wooden fish trap built by pushing narrow splints cut from a young pine tree to the bottom of the pond into a fish-trap-shaped snare. Source: Photo Archives of the Finnish Defence Forces.

of the occupation, there were plenty of independent professional fishermen working in the area. The fishermen were allowed to take a specified portion of the catch for their own needs, but the largest part of the catch was to be sold to the fish receiving depots run by the military administration, so the fish would be available for army and general consumption. These depots had their own icehouses, salting houses and smokehouses, which were in continuous use during the fishing season.[39]

More specific directives on fishing in East Karelia were not given until the spring of 1943.[40] For the most part, the regulations were in accordance with the fishing laws in force in the Finnish homeland, defining, for instance, the

39. Siltamaa 1984, p. 51.

40. Edict on fishing in the area of the East Karelia military administration no 18. Itä-Karjalan sotilashallinnon säädöskokoelma nos 18–20, 1943.

size of net eyelets, minimum fish size by species and closed season for crayfish. In addition, fishing licences, which were required for professional fishing, were regulated. The military units were allowed to organise independent fishing to meet the needs of their men without special permission.[41] Salted, smoked and fresh fish brought much needed variety to the soldiers' diets. In addition, fishing – just like hunting – helped individual soldiers and officers to relax somewhat in nature in the middle of war.[42]

Small and big catches

The military organisations had their own fishing officers, who were responsible not only for organising fishing activities and procuring nets for the fishermen, but also gathering information on fishing waters, and preparing statistics. Second Lieutenant Erkki Halme acted as the fishing officer for one unit and later, as a well-known fisherman, a researcher and professor in the field of ichthyology. The duties of the fishing officer also included inspecting the fishing detachments. It becomes clear from his inspection reports that the fishermen were not in all cases so enthusiastic about fishing.

On one inspection tour, Halme noted that when he arrived at the detachment's bunkers in daytime, all twelve fishermen were inside and most were not engaged in any activity. In another fishing detachment he found 200 nearly untouched nets.[43] Area headquarters later sent an order to the detachments stating, 'if the daily catch of the fishing detachment's fishing group is under two kilograms per man per day, there is no call to keep the group fishing'.[44] The norms familiar from the Soviet economy were in this way adopted into the economy of the Finnish military.

But usually fishing gear was used and often the fishing yielded well. The previously mentioned Erkki Halme remembered one of his platoons pulling in a catch of biblical proportions in the early spring of 1943:

> The ice hole was in the middle of the lake. The men's legs were up to their calves in water. They were doubtful as to whether the ice would take the weight of the entire haul. With the support of heavy ropes and long poles, they succeeded in pulling the seine onto the ice. A truck and ten horse-drawn sleds were needed to

41. Report of the fishing officer, Archives of the Maaselkä group headquarters, SA.
42. Interviews with veterans of the 14th division.
43. Report of the fishing officer, Archives of the Aunus group headquarters, SA.
44. Halme, 'Report of the head of the fishing company', Archives of the Maaselkä group headquarters, SA.

take the catch to the city. The weigh-in at the food and feed magazine indicated 12,000 kilograms of bream and 200 kilograms of pike-perch.

In 1943, the fishermen brought in over 535,000 kilograms of fish to the fish receiving depots, of which 115,000 kg remained for the use of the fishermen themselves.[45] When gauging the size of the entire take, we need to add the amounts caught by the units and other fishermen to these figures; however, these have not been estimated. These wartime catches from East Karelia appear significant. But in reality, the fish from East Karelia were only a fraction of the national fish catch of Finland for 1943, which was about 28 million kilograms.[46]

Fishing was, however, important for soldiers and civilians living in East Karelia. One consequence of this was that, by the end of World War Two, East Karelia's lakes were the most thoroughly known waters in wartime Finland. The reason for this perhaps slightly surprising situation was that the Finns in late 1941 took control of the Russians' Kentjärvi biological field station, run by the Leningrad Society for Naturalists.[47] This lake-region station, situated in what had been the USSR, became the University of Helsinki's first field station for freshwater studies. The finest Finnish professors of animal and plant sciences and geography of their day started to work in summer 1942 at this scientific station beautifully built of logs.[48] An East Karelian village thus became the centre of empirical natural scientific studies in wartime Finland.

Basic scientific research on fish and game was undertaken in order to develop the nature-based livelihoods of East Karelia and to harness the territory's natural resources for the use of the future state of Greater Finland.[49] In order to organise fishing activities, the fishing officers used printed questionnaires to gather precise information on East Karelian fishing waters from professional fishermen and others who used the waters. The research results produced by Russian ichthyologists in the inter-war period at the Kentjärvi biological station were taken advantage of too. Data was gathered on each lake: size, characteristics, time of freezing, and key fish species and their populations and spawning

45. Siltamaa 1984, p. 50.

46. Data provided by the Finnish Game and Fisheries Research Institute in 2020.

47. Today the name of the Kentjärvi biological station is Kontsezerskaya biologitseskaya stantsiya.

48. Report on the activities of the University of Helsinki, academic year 1941–42, p. 58: Report on the activities of the University of Helsinki, academic year 1942–1943, p. 72; Central Archives of the University of Helsinki, university senate meeting minutes, card 1/11, 7 Jan. 1942, 8 § and card 2/14, 10 Feb. 1943, 23 §.

49. Laine 1993, pp. 184–85.

Figure 5. The University of Helsinki's first field station specialising in biological research of freshwaters operated in East Karelia in 1941–1944. Preceding the biologic station was the Borodino limnological research station founded by the Russians in 1897. The station was moved to the Karelian village of Kentjärvi by the end of the 1920s. After the war, the station was moved to the control of the University of Petrozavodsk. Source: Photo Archives of the Finnish Defence Forces.

seasons. Detailed maps indicating the lake depth, bottom quality, fishing spots for different kinds of equipment and fish species, and the quality of the fishing spot (good, satisfactory, poor) were prepared for the ten largest lakes.[50] Dr Erkki Halme collected the data on the limnology and ichthyology of the East Karelian lakes into a manuscript hundreds of pages long. It was, however, not published as a book, since the area reverted to the Soviet Union at the end of the war. Nevertheless, no equally broad scientific-economic study of Finnish lakes has ever been conducted.

In addition to the lakes of East Karelia, the fisheries management administration planned on tapping the resources of the White Sea and the Barents Sea, which belonged to the Soviet Union. The White Sea contained abundant stocks of oceanic bivalves, crabs, fish species, seals, walrus and whales that Finns were not very familiar with. The Finns were interested, for instance, in the

50. Erkki Halme, archives of the Aunus group headquarters; Heikki Järnefelt, Archives of the East Karelian military administration scientific unit, SA.

plentiful harp seal, whose total population in the Barents Sea was estimated at 10–11 million individuals. The beluga whale was also of interest to the military administration, but surprising hurdles to hunting the species arose: 'Hunting it would be extremely productive, but the local inhabitants have been very resistant, as they consider it a holy animal.' According to the locals living on the shores of the White Sea, beluga whales drove fish shoreward from the open water, from where inhabitants could then catch them.[51] The cultural heritage of local people based on long-term observation of local coastal environs apparently protected valuable natural resources.

The effects of hunting and fishing

Fishing and hunting in East Karelia were monitored from time to time and punishments were even meted out. The use of explosives for fishing was considered particularly reprehensible, and there are records of punishments in the units' orders of the day.[52] Even during the retreat, those who had fished using the 'pioneer method' were threatened with court martial. However, despite the fact that there were attempts to closely regulate hunting and fishing, soldiers exhibited lapses in following many of the regulations.[53] This was the case in East Karelia in 1941–1944 as well. Hunting seasons were ignored and game was shot on sight, regardless of season or species. This was the case particularly in the vicinity of the front lines, although further back there were attempts to better monitor the edicts.[54]

There were attempts to enlighten and educate the soldiers,[55] but their effects remained minor. Since hunting and fishing violations had been dealt with lightly in the Finnish homeland, respect for such regulations in conditions of war was even more irregular. Furthermore, the soldiers were in a foreign land, perhaps only temporarily, and this also affected the morale of the hunters and fishers negatively. In East Karelia, even high-level officers were guilty of violations. One well-known general asked about moose when visiting a wilderness

51. Archives of the Kalastusmuseosäätiö, MMM, MH, Tiedusteluosaston kirjelmiä, Kalataloudellinen katsaus; Maantieteellis-hydrologinen selostus, Captain H. Järnefelt, 12 May 1942; Information on the fishing on the Muurmansk shores and the Barents Sea, Captain H. Järnefelt, 1 June 1942.
52. Infantry Regiment (JR) 52 order of the day, 17 Nov. 1941, SA.
53. Soikkanen 1999, p. 152.
54. Ollikainen 1985, p. 90; Interviews with veterans of the 14th division.
55. See Kivilinna 1944; Halme, Memo to units about the harmfulness of the use of explosives in fishing, 12 May 1942. Archives of the Aunus group staff, SA.

base. The soldiers informed him that they existed in the vicinity, noting that it was only allowed to hunt moose during the official hunting season. To this the general responded: 'If there's a lack of food, go ahead and shoot and say it stepped on a mine.'[56] Since the upper military leadership turned a blind eye to infractions, it was difficult to question illegal hunting by the enlisted men.

Figure 6. A capercaillie that stepped on a land mine? According to the descriptive text about this photograph taken from the front, 'a handsome capercaillie has stepped on a land mine' in no man's land. The surroundings hinting of an early spring combined with the immaculate shape of the bird, however, give credence to the assumption that it's been illegally hunted during courtship. The conditions during the war saw a general decline in law-abidingness.
Source: Photo Archives of the Finnish Defence Forces, photographer Tk-Holming.

But what sort of effect did this apparent plundering have on the fish and game populations of East Karelia? A unique study on the ecological ef-

56. Tapola 2004, pp. 164–65.

fects of the war is zoologist Teppo Lampio's survey of game stocks conducted in 1941–1944.[57] His study gives a precise picture of the game in the area, and also partially of the effects of the war on game populations. One development is convincingly demonstrated in the study. When East Karelia was occupied, game was abundant, but by the time Finland retreated, game had become scarce. Lampio categorised the abundance of stocks on a scale from 5 to 1, in other words from very abundant (5) to very scarce (1). The population of only one species, the mallard, was approximately at the same level when Finland left East Karelia as it was when it entered, in this case averagely abundant. The other populations had decreased dramatically. By the spring of 1944, moose had decreased in the territory from very abundant to very scarce, and things had gone even worse for the black grouse. The capercaillie had also dropped from abundant to scarce. The squirrel had diminished to non-existent, but the researcher noted disease and lack of food in addition to hunting as reasons for its demise. Even though game was susceptible to the destruction and disturbance caused by war, Lampio's study convincingly indicated that the main reason for the severe drop in amounts of game was the merciless hunting practised by the soldiers and officers on both sides of the front.

As far as fishing is concerned, there are no Finnish studies of fish species in the area, but research was conducted on the fish stocks in waters and how they could be exploited in fishing and feeding the troops. There is also a lack of detailed fishing statistics, since, due to the hurried evacuation of the military administration of East Karelia in June of 1944, 'a worthless section of the archives was burned to ease transport difficulties', at which time the greater part of the fisheries administration archives were destroyed.[58] In all probability, fish stocks did not, however, suffer nearly as much from warfare as game populations. Even though explosives were used, many waters suffered from overpopulation and stunted fish, in which case the explosive hunting may have served to thin out overly dense populations and improve the conditions of those individuals that survived. Furthermore, the wilderness warfare of the time did not cause the same kind of environmental damage to fishing waters that contemporary armies and warfare can cause.

57. Lampio 1946, pp. 19–31.

58. Syrjö 1972, p. 318.

	1941	1942	1943	1944
Brown bear	2	1	2	2
Fox	5	4	4	4
European otter	1	1	1	1
Stoat	2	2	2	2
Least weasel	?	?	1	?
Wolverine	1	2	2	?
Lynx	1	1	1	?
Brown hare	4	4	2	2
Red squirrel	5	5	3	1
Moose	4	4	2	1
Wild forest reindeer	2	2	?	?
Mallard	3	3	3	3
Common teal	2	2	2	2
Wigeon	1	1	1	1
Common goldeneye	3	3	3	3
Red-breasted merganser	2	2	2	2
Loon	2	2	2	2
Common snipe	1	1	1	1
Willow grouse	3	3	2	2
Black grouse	5	4	2	1
Wood grouse	4	4	2	2
Hazel grouse	4	4	3	3

Table 1. A unique collection of research material gathered over the course of the war by a zoologist Teppo Lampio displayed the ruthless nature of hunting in the East Karelian region. Before the arrival of the Finnish troops, the area had abundant game (5/4), but only a few animals remained after the troops' departure (2/1). (Data Source: Lampio 1946, 21.)

East Karelia: The fish and game store of Finland?

In the collective Finnish mindscape, East Karelia was made into the land of poetry, song and music and seen as belonging to the Finnish state. But when Finnish control of the mythical East Karelia was finally achieved during World War Two, Finland entered it by violence, through warfare and as a conqueror.

For over a little less than three years, East Karelia was governed by a Finnish military administration that attempted to regulate fishing and hunting by applying legislation from peacetime Finland. This strategy was an unrealistic choice in conditions of war, which favoured illegal practices. Despite this, we can say that there was a clear principle of sustainable use of natural resources evident in the policies of the military administration. Also, the Finnish army strove to take conservation of rare species such as forest reindeer into consideration as early as World War Two. So-called Khaki conservation – that is, attempts by the military to protect nature – is often thought to have started in the late twentieth century. In this current framework, the wartime initiative of the Finnish army seems to provide a particularly early example.

Hunting and fishing offered Finnish soldiers exceptional opportunities to relax in nature in wartime conditions. However, the hunting lands and fishing waters of East Karelia did not meet the ambitious goals set for them by military administration. The region did not become the inexhaustible store of game and fish that would supply the 'bread of the waters' for the populace of the Finnish homeland or enough furs for export. Fishing did generate, however, a significant dietary supplement both for the troops as well as the local populace, and there was something left over to send back to the home front. The effective organisation of fishing and its productivity were, indeed, significant factors in the wartime food administration of East Karelia.

Hunting was of less significance. Even during peacetime conditions, hunting in East Karelia had been significant primarily in terms of fur-bearing game. Following collectivism and the Stalinist persecutions, the significance of hunting diminished even further. During World War Two, the game of East Karelia did offer extra income and, in certain areas, a significant supplement to the rations of Finnish soldiers. The downside was the destructive effect of hunting on the game stocks in the area.

The implications of this study are obvious. The war had a devastating effect on wildlife on the frontlines. However, this conclusion raises an important follow-up question. How did the transfer of about half a million able-bodied Finnish men to the eastern front affect fishing and hunting, and thus the game and fish stocks of the home front, which made up the overwhelming majority of the country's area? It seems that this perspective has not been studied in any systematic manner so far. Nevertheless, there are strong indications that the significant reduction in hunting and fishing behind the front had a very positive effect on the number of seals, large birds of prey and large carnivores in wartime Finland. In conclusion, in order to answer such important questions about the overall environmental effects of wars, we need comprehensive long-term studies of entire countries in the future. Yet, we should be careful not to greenwash warfare.

Bibliography

Archives

Archives of the Finnish Literature Society, Helsinki, Finland.

Archives of the Fishing Museum Association, Riihimäki, Finland.

Central Archives of the University of Helsinki, Helsinki, Finland.

The National Archives, Helsinki, Finland.

Literature

Airaksinen, K. 1943. 'Metsästys Itä-Karjalassa neuvostovallan aikana', *Metsästys ja Kalastus* 2: 39–43.

Autio, Sari. 2002. *Suunnitelmatalous Neuvosto-Karjalassa 1928–1941*. Helsinki: Suomalaisen Kirjallisuuden Seura.

Henttonen, Antti and Matti Lappalainen. 1991. *Tuolla puolen Syvärin. Rajajääkäripataljoona jatkosodassa*. Keuruu: Otava, 1991.

Halme, Erkki. 1943. 'Itä-Karjalan kalavesistö', *Luonnon Ystävä* 4: 98–100.

Holm, Paul. 2012. 'World War II and the "Great Acceleration" of North Atlantic fisheries'. In Simo Laakkonen and Richard Tucker (eds). 'World War II, the Cold War, and Natural Resources' Special Issue of *Global Environment* 10 (o.s.): 66–91.

Hämynen, Tapio. 1998. 'Karjalan yhteiskunta ja talous 1800-luvun lopulta toiseen maailmansotaan'. In Pekka Nevalainen and Hannes Sihvo (eds). *Karjala. Historia, kansa ja kulttuuri [Karelia: History, People, and Culture*]. Helsinki: Suomalaisen Kirjallisuuden Seura.

Kairikko, Juha K. 1991. *Seitsemän vuosikymmentä metsästykselle. Suomen Metsästäjäliitto - Finlands. Jägarförbund ry:n 70-vuotishistoriikki*. Hanko: Suomen Metsästäjäliitto.

Kangaspuro, Markku. 2000. *Neuvosto-Karjalan taistelu itsehallinnosta. Nationalismi ja suomalaiset punaiset Neuvostoliiton vallankäytössä vuosina 1920–1939*. Helsinki: Suomalaisen Kirjallisuuden Seura.

Kivilinna, M. 1944. *Sotilaan metsästysopas*. Helsinki.

Kulha, Keijo K. 2004. *Mies ja vuosisata. Pääkonsuli Aarne Koskelon tarina 1900–1998*. Jyväskylä: Gummerus.

Laakkonen, Simo and Timo Vuorisalo (eds). 2007. *Sodan ekologia. Nykyaikaisen sodankäynnin ympäristöhistoriaa*. Helsinki: Suomalaisen Kirjallisuuden Seura.

Laakkonen, Simo, Richard Tucker and Timo Vuorisalo (eds). 2017. *The Long Shadows: A Global Environmental History of the Second World War*. Corvallis: Oregon State University Press.

Laakkonen, Simo, J.R. McNeill, Richard P. Tucker and Timo Vuorisalo (eds). 2019. *The Resilient City in World War II: Urban Environmental Histories*. New York, London: Palgrave Macmillan.

Laakkonen, Simo and Alla Bolotova. 2021. 'Ristiaallokossa: Laatokan pilaantumisen ja suojelun ympäristöhistoriaa 1800-luvun lopulta 2000-luvulle'. In Maria Lähteenmäki (ed.) *Laatokka. Suurjärven kiehtova rantahistoria.* Helsinki: Suomalaisen Kirjallisuuden Seura. pp. 131–61.

Laine, Antti. 1993. *Tiedemiesten Suur-Suomi. Itä-Karjalan tutkimus jatkosodan vuosina.* Helsinki: Historiallisia arkistoja 102.

Lampio, Teppo. 1946. *Havaintoja Itä-Karjalan riistaeläimistöstä,* Suomalaisen Eläin- ja Kasvitieteellisen Seuran Vanamon eläintieteellisiä julkaisuja, osa 1.

Lehikoinen, Heikki. 2007. *Tuo hiisi hirviäsi. Metsästyksen kulttuurihistoria Suomessa.* Helsinki: Teos.

Nieminen, Markku. 1998. 'Vienan Karjala'. In Pekka Nevalainen and Hannes Sihvo (eds). *Karjala. Historia, kansa ja kulttuuri.* Helsinki: Suomalaisen Kirjallisuuden Seura.

Nygård, Toivo. 1978. *Suur-Suomi vai lähiheimolaisten auttaminen. Aatteellinen heimotyö itsenäisessä Suomessa.* Helsinki: Otava.

Näre, Sari and Jenni Kirves (eds). 2014. *Luvattu maa. Suur-Suomen unelma ja unohdus.* Helsinki: Johnny Kniga.

Ollikainen, Jorma. 1985. *Pieninkä. Erätarinoita Itä-Karjalasta 1942–1944.* Kuopio: Kustannuskiila.

Pohja-Mykrä, Mari, Timo Vuorisalo and Sakari Mykrä. 2005. 'Hunting bounties as a key measure of historical wildlife management and game conservation: Finnish bounty schemes 1647–1975'. *Oryx* **39** (3): 284–91.

Robertson, Thomas, Richard Tucker, Nicholas Breyfogle and Peter Mansoor (eds). 2020. *Nature at War: American Environments and World War Two.* Cambridge: Cambridge University Press.

Siltamaa, Erkki. 1984. 'Muistelmia Itä-Karjalan kalastuksesta 1940-luvun alussa'. *Suomen Kalastuslehti* 2: 50.

Simonen, Seppo. 1971. *Vako Oy. Kaupallista toimintaa Itä-Karjalassa 1941–1944.* Helsinki.

Soikkanen, Mauri. 1999. *Sotilaamme erämiehinä Itä-Karjalassa 1941–1944.* Jyväskylä: Gummerus.

Soikkeli, Martti. 2005. 'Hirvenmetsästystä hävittäjäkoneella Vienan Karjalassa'. *Parivartio,* 29 April.

Sopanen, Valio. 1992. 'Äänisjärvellä herkuteltiin kaloilla ja kanoilla'. In Raine Valleala (ed.) *Kuusaalaisia jatkosodassa. Peikkopataljoonan tie.* Kuusankoski.

Syrjö, Veli-Matti. 1972. 'Itä-Karjalan sotilashallinto ja sen arkistot'. *Historiallinen Aikakauskirja* 4.

Tapola, Päivi. 2004. *Ajan paino. Jalkaväenkenraali K. A. Tapolan elämä.* Helsinki: Tammi.

Ylänne, Yrjö. 1941. 'Villipeuran puolesta'. *Metsästys ja Kalastus* 10: 257–58.

SECTION 3

ALTERING THE ENVIRONMENT

Chapter 10

FROM STALE AIR TO TOXIC: CONCERNS ABOUT URBAN AIR IN FINLAND

Janne Mäkiranta

For one week in May 1969, air pollution made the headlines in the Finnish press, as central Helsinki was dominated by a variety of protests about the unacceptable state of the city air. In the central square stood three-metre-high plastic lungs covered with paste that gathered soot and dust from the air. The student demonstrators entered public transport in gas masks and showed posters with slogans such as 'Lead kills slowly!' and 'Lead and sulphur explode our organs!' Even a choir was gathered in the central square singing about the dangers of strontium, a substance known to the public from the nuclear fallouts of the 1950s. Next to the choir was an oxygen bar where fresh air and information about the effects of air pollution were provided to passersby. The most sensational form of protest was the ceremonial destruction of a car, the most emotive enemy of urban air.[1] The Pollution Week demonstration, as it was labelled, was organised by students of veterinary medicine from the University of Helsinki. It was a manifestation of resentment towards urban air quality in the Finnish capital in the late 1960s.[2] This resentment was part of a wider phenomenon, in which the potential health effects of air pollution caused increasing concern in many European countries and in the United States. The Pollution Week concept itself was imitated from the United States. Thus, following a transnational trend, urban air pollution had become a topical environmental concern in Finland.

The Pollution Week demonstration was the first public protest against the misuse of the environment in Finland. Despite this, air pollution has received relatively little attention in Finnish environmental history. Historical works on public health and urban environment have examined changes in air quality in a

1. 'Kaasunaamariajelu ympäri Helsinkiä', *Helsingin Sanomat*, 24 April 1969.
2. Schönach 2008, pp. 17–80.

doi: 10.3197/63824846758018.ch10

few larger Finnish towns. They have shown how urban air quality represented a concrete and immediate environmental irritation for ordinary people.[3] However, it can be argued that forests, lakes and coastal waters form the essential arenas through which Finnish environmental history is depicted. Confined to a few major cities and industrial areas in a sparsely populated country, air pollution problems appear limited compared to the simultaneous degradation of forests and waters. The purpose of this chapter is to show, however, that concerns about air quality form an essential aspect of people's relationship towards the environment. Despite a low population density and relatively small industrial sector, concerns for clean air in Finland followed transnational trends that began with the hygiene movement in the early twentieth century. After the Second World War scientific analysis of the hazards of urban air became the primary way of evaluating urban air quality, paving the way for more nuanced concern about toxic substances in air. At the same time, the concern about urban air in the late 1960s was in many ways a continuation of older concerns about an unhealthy environment and the deleterious effects of civilisation.

Hygiene and the benefits of clean air

In the 1860s, Finnish botanist William Nylander made an observation that would later be regarded as pioneering in air pollution research. Although he was the professor of botany at the University of Helsinki, Nylander subsequently emigrated to France where he conducted his most important studies and gathered extensive collections of plants. Nylander's most important point of interest was lichen, the somewhat mysterious plant that seemed to be an ill fit in the botanical taxonomy. At the height of his scientific career in the 1860s, Nylander surveyed the distribution of different species of lichen in the streets of Paris. He observed that most species of lichen had degraded or vanished altogether from the urban areas, while they still thrived in parks and the outskirts of the city. This led Nylander to conclude that the air in urban areas had become unfit for lichen. Nylander proposed that lichen could be used as a sort of 'hygiomètre' that could measure the healthiness of air. The lichen showed, for example, that people in Paris should take their children to enjoy the air in the park of Luxembourg rather than in the streets of the city.[4] Nylander later made similar observations in Helsinki, indicating that the small northern town had similar issues with air quality. Nylander's views later became an example of

3. Kruut 1999; Harjula 2003; Schönach 2008; Mäkiranta 2022.

4. Nylander 1866, pp. 365–71.

an early warning about urban air pollution. He was cited by those who began to study urban air quality in Finland in the 1950s, and also by some Finnish historical studies on air pollution.[5]

However, a closer look at Nylander's study indicates that he was not particularly concerned about substances such as smoke and soot that are usually related to air pollution in industrialised cities. In fact, when one of Nylander's French colleagues proposed that the death of lichen could be caused by black soot from factory chimneys Nylander hesitated and pointed out that many species of lichen have black pigmentation naturally. Instead of smoke and soot, Nylander refers to impure and 'imprisoned' air in densely built environments. Lichens could not survive in these environments, Nylander explained, because they depended on the free circulation of air.[6] Rather than being an early warning about the health effects of air pollution that caused concern in the 1950s, Nylander's hygiomètre represents the mid-nineteenth century concern in Europe about stale air in the densely built and congested cities.

In the mid-nineteenth century, the prevailing hygienic thought saw cramped, dirty and stuffy urban environments as unhealthy and this view was backed up by the novel public health statistics. As Charles Rosenberg has argued, the idea that the urban environment was unhealthy compared to the countryside can be seen as an intuitive truth in the civilisation critique of the nineteenth century. The stuffy and dirty cities manifested the ill effects of progress and civilisation, while fresh air in the countryside provided a healthy sanctuary.[7] Although miasma theory as such was losing its place in hygienic thought, the importance of clean air retained its significance in the early twentieth century, both in Europe and in the United States.[8] In Finland, clean air became one of the essential features of the health education promoted by public health experts at the turn of the century.[9] Ideas of clean and fresh air persisted among Finnish medical experts even alongside germ theory, as air was one factor affecting the so-called susceptibility of individuals, which germ theory alone could not explain.[10] Thus, in early twentieth century Finland, air was an important aspect of environmental health grounded in the hygienic thought.

5. Hosiasluoma 2001, p. 196; Schönach 2008, p. 46.
6. Nylander 1866, pp. 365–71
7. Rosenberg 1998, p. 720.
8. Thorsheim 2006, p. 39.
9. Lehtonen 1995; Saarikangas 1998; Jauho 2007.
10. Jauho 2007, p. 95.

Failures of smoke abatement

The importance of clean air did not, however, translate into effective measures against polluted air. Although the scale of problems was different in Finland compared to industrial centres in Europe and the United States, Finnish urban dwellers were not spared the ill effects of modern society on their ambient air. Smoke from heating furnaces, trash incinerators, factories and to a growing extent automobiles caused nuisance to people living in urban centres such as Helsinki and Tampere.[11] The foul-smelling pulp industry formed the most widespread air pollution issue in rural industrial communities. Since the First World War, more diverse industrial activities were increasingly introduced to Finland, which also diversified the modes of air pollution. The copper industry in the small town of Harjavalta became the most infamous example of this rural air pollution that could destroy forests twenty kilometres from the factory.[12] To remedy the nuisance and to safeguard the healthiness of urban air, similar measures of abatement were adopted as were used widely in Europe. The primary forum was the municipal health boards established in the late nineteenth century following the British example. A smoke inspector was appointed in Helsinki in 1901, a system copied from Munich, but the office lasted only for a few years. Nuisance laws were also passed in the 1920s in order to curb foul smells from the pulp industry, but these laws were never used against major industries.[13]

In Finnish historical works on air pollution, the failure of air pollution control in the second half of the twentieth century has usually been seen to lie in an inability to prove that smoke and dust were dangerous to health and in the dominance of economic interests.[14] Both these arguments have their merits. The importance of air in hygienic thought was grounded more on vague ideas of healthy fresh air rather than any specific ill effects of smoke. As Finnish hygienic education shows, the direct health effects of these substances were regarded as questionable well into the mid-twentieth century.[15] It has even been argued that, in the smoke abatement movements in Europe and the United States, economic and aesthetic reasons came well before any health concerns.[16] Similarly, when Finnish industries adopted cleaning techniques from the United States in the

11. Schönach 2008; Harjula 2003.
12. Mäkiranta 2022, pp. 53–55.
13. Harjula 2003, p. 181.
14. Ibid.; Schönach 2008, p. 95; Lahtinen and Vuorisalo 2004, pp. 685–90.
15. Mäkiranta 2022, pp. 34–37.
16. Uekötter 2009, p. 44.

early twentieth century, it was mainly to avoid compensation demands from landowners and to prevent material waste.[17]

It could be argued that the hygienic concern for clean air was too vague and not powerful enough to instil real action that would have economic consequences. However, despite its vagueness, the hygienic idea was an essential aspect in the debate about urban space that was increasingly prominent in the twentieth century. Robert Jütte has shown how the complex questions about how cities should smell, look or sound were being renegotiated in the early decades of the twentieth century.[18] In Finland a turning point in this regard came in the late 1950s. As the hygienic concern about clean air challenged the state of affairs, officials responded in a way that would change not only the way air quality was managed but also the underlying concern about air quality.

Answering concerns with research

In the late 1950s, the real estate committee of the city of Helsinki received a complaint about dead trees in the district of North-Haaga. It was presumed that these trees had been killed by polluted air.[19] There was nothing new in the fact that trees and vegetation suffered from air pollution in urban and industrial surroundings. In the late 1940s and early 1950s, for example, Finnish chemical and metal industries had caused massive damage to forests in industrial towns. Local inhabitants had also been concerned over the potential effects on health, drawing similar conclusions to those of Nylander. Knowledge about air pollution had become more specific since the days of Nylander, however. Finnish medical experts were fairly certain that, although pollutants such as sulphur dioxide could kill trees, they were rather harmless to humans in the same amounts.[20] The link between plant life and healthy air did not hold such sway as it had done for Nylander.

However, the hygienic importance of clean air had not vanished from people's ideas of a healthy environment. While the complaints in small industrial towns amounted to little, the death of trees in the leafy North-Haaga was a different matter. This was a newly built neighbourhood surrounded by woods, inhabited by the health conscious urban middle class, and specifically designed to exclude the unwanted urban environmental elements that were seen as det-

17. Noro 1958, pp. 234–44.
18. Jütte 2018.
19. Record of the Realstate Committee, 31 March 1958, Helsinki City Archives.
20. Noro 1958, pp. 234–44.

rimental to health.[21] This middle-class enclave challenged the idea of what was the acceptable state of urban air. In a way, this serves as the Finnish equivalent of the well-known significance of middle-class resistance in twentieth century environmental history.[22] As Simo Laakkonen and Timo Vuorisalo have argued, members of the working class suffered the most from environmental pollution, but they also developed a somewhat fatalistic attitude on the issue that differed from that of the middle class.[23] Studies on Finnish industrial towns have also shown that the inhabitants valued local industries greatly and did not always make a noise even about clear hazards for health.[24]

Due to the complaints, the officials in Helsinki took action and they did it in a new way. Whereas complaints about air quality had previously been handed over to the city hygienist and public health board for evaluation, this time the committee decided to order a full investigation on the air quality of Helsinki. The Finnish Institute of Occupational Health(FIOH) was chosen to conduct the study. The study was applauded and closely covered by the press, with excited descriptions of the measuring apparatuses and their use.[25] Despite the attention it received, the study was not particularly important in itself. The results did not reveal anything special or alarming about the air quality of Helsinki and did not spur any immediate action. To be precise, it was not even the first investigation on air quality in Helsinki. It was actually preceded by a more limited and little noticed study by the Helsinki School of Technology a few years earlier.[26]

What is important in the FIOH investigation of 1959, however, is the fundamental change in urban air quality management that it depicts: the idea that scientific investigations on air quality could help to solve the problem. As Frank Uekötter has argued, this was a revolutionary change that took place first in the United States after the Second World War. Studies on air quality had been conducted before in many countries, but they had been sporadic. Systematic measurements were considered rather unnecessary, even by the most ardent anti-smoke activists.[27] The situation changed in United States from the late 1940s onward, as increasing complaints about urban air quality

21. Kolbe 2007, pp. 32–38.

22. See, for example, Sellers 2012.

23. Laakkonen and Vuorisalo 2019, pp. 288–90.

24. Lahti and Saarela, p. 309; Ahlberg, p. 61.

25. Schönach 2008, p. 152.

26. Kajanne and Laiho 1958.

27. Uekötter 2009, p. 129.

were answered with scientific research, which it was hoped would produce rational and objective solutions to the complex problem. The investigation on air quality in Helsinki, ordered by city officials, was the first indication that a similar faith in scientific expertise on air pollution was taking root in Finland.

After the FIOH study, the importance of research was further highlighted by the Smell Nuisance Committee established in Helsinki in 1960 to deal with complaints about trash incinerators. The committee applauded the study made in 1959 and urged municipal officials to order more studies from FIOH on air quality.[28] In 1965, a government expert advisory body on air pollution issues, led by FIOH personnel, was created. This cemented air quality research and expertise at the national level. The necessity of research was taking root outside Helsinki too. During the latter half of the decade, the FIOH conducted preliminary investigations in several towns and industrial areas.[29]

The pathogens in air

This new air pollution research had a somewhat different take on air compared to the older hygienic view. Methods and principles to study the health effects of urban air had roots in industrial toxicology and industrial hygiene, which had been focusing on non-microbial disease agents since the late nineteenth century. As Christopher Sellers has shown, it was in these disciplines that different dust and fumes were established in the early twentieth century as specific causes of disease. By the logic of 'the dose makes the poison' industrial toxicology could, at least in theory, deduce what kind of levels of different substances were harmful for health.[30] Armed with this kind of knowledge, industrial hygiene could then turn the situation around, and monitor the air inside factories in order to provide a safe working environment.

This industrial toxicological logic was adopted outside of factories when the management of environmental pollution became increasingly pertinent in the twentieth century. In Finland, the intimate connection between industrial environment and environment in general was epitomised by the fact that it was FIOH that became the national author in air pollution research, having

---

28. Report of the Smell Nuisance Committee to Helsinki City Council 23 Sept. 1966. Archives of the FIOH.

29. Studies were made in Turku (1966), Lappeenranta-Lauritsala (1967), Pori (1967–68), Kokkola (1966–67 and 1970–71), Rauma (1971), Tampere (1971), Valkeakoski (1971–72), Kuusankoski (1971–72), Karhula (1971), Mänttä (1971–72).

30. Sellers 1997.

first adopted the practice of industrial hygiene from United States.[31] Environmental historians have criticised the spread of industrial toxicology into other areas of environmental control. It has been seen as a reductionist approach that fails to consider the intricacies of environment and its relationship to health.[32] Linda Nash has argued, for example, that despite its interest in environmental pollution, post-World War Two toxicological research did not provide a return to the holistic relationship between human health and environment that was once a prevailing feature of medicine and still retained its place in lay people's views. On the contrary, research on environmental toxins reduced the question into measurement of quantity.[33]

Indeed it can be said that, from the point of view of air pollution research, the once important concept of clean air did not exist at all. Air was always a mixture of substances both natural and man-made. Since everything in air that deviated from the basic atmosphere could be considered impurity, the concept of clean air had little practical value in air pollution research. The essence of air pollution research was to analyse the different substances in air in order to determine which were harmful and in what quantities. In other words, the research was focused on the specific impurities in air and their specific effects. This was a marked shift from the vague but unquestioned hygienic idea about the benefits of clean and fresh air.

On the other hand, by framing the air quality issue as a toxicological puzzle, research on the health effects of air pollution specified the ill effects of polluted air into specific pathogens in air. To examine the specific effects of each substance in air became the overwhelming task of this new expertise. With this puzzle came also an elevated concern for the chronic effects of long-term exposure to polluted air. When the first international conferences on air pollution research were held in the early 1950s United States, many attendants saw the potential chronic effects as much more dangerous than occasional smog disasters in heavily polluted areas.[34] Chronic poisonings were known from occupational medicine, but little was known about the threat posed by normal urban air. The nascent threat of carcinogens was a further concern, about which little was known.[35] Thus, research on the health effects of air pollution

31. Mäkiranta 2022, pp. 55–59.

32. Nash 2008, pp. 650–55; Vaupel and Hombur 2019, p. 37; Jas 2015, pp. 52–53.

33. Nash 2008, pp. 650–55.

34. Kehoe 1952, p. 477; Phair 1956, pp. 3–10.

35. Hueper 1952, p. 30.

changed the vague notion of bad air into specific puzzle about the health effects of different substances.

There was, however, an epistemological off-balance in the methods to solve this puzzle. Even with the less than state-of-the-art equipment and laboratories of the FIOH, it was possible to analyse dozens of substances from Finnish urban air. Due to advances in analytical chemistry, even the elusive polycyclic hydrocarbons could be revealed in routine studies. The significance of these substances for public health was a more ambiguous matter. Although more sophisticated epidemiological studies had strengthened the link between urban air quality and various chronic respiratory diseases such as chronic bronchitis and lung cancer, the causal links between different substances and their effects were far from evident in the mid-1960s.[36] This had a marked effect on how the results of studies presented the issue. For example, after showing the multitude of substances, including carcinogens, in Finnish urban air, the FIOH's studies could only state that the quantities were in all likelihood too small to cause any danger to health.[37]

The new expertise on air pollution changed the way polluted air was depicted from vague notions of smoke and staleness to specific chemical substances. But the way the FIOH experts viewed the uncertainties and unknowns in the issue was not well received by the public. The first study in 1959 had been a novelty, but in the 1960s these air quality studies were increasingly criticised for ambiguous results and lack of real practical recommendations.[38] At the same time, attitudes against air pollution hardened, especially in Helsinki, where air quality decreased in the late 1960s due to the growing number of private cars. When this indignation was coupled with the rising environmentalist critique against the deleterious effects of industrial society in general, the toxicological view on air quality became the core of the new concern about urban air.

The environmentalist concern

Although Rachel Carson's *Silent Spring* is often touted as the herald of environmentalist thinking, many Finnish environmental historians have argued against this idea in the case of Finland. It has been shown, that despite its translation into Finnish in the early 1960s, the book provoked relatively little conversation and concern over pesticides was not regarded as important in Finland. Instead,

36. Heiman 1967, pp. 488–99.

37. Jormalainen, Laamanen and Lehtinen 1961, p. 152; Laamanen and Noro 1967, p. 16.

38. Schönach 2008, pp. 187–91.

the rising concern about environmental degradation was connected to multiple causes, both domestic and transnational. These developments created a situation in which people were concerned for their own immediate environment but at the same time linked these events into global ones that were made visible to the Finnish audience by newspapers and television.[39] The concern over urban air pollution appears to follow the same formula. The indignation about local ambient air quality was connected to urban air pollution problems in other countries and the pollution of the environment in general. This view is most evident in the pamphlet literature about environmental pollution that begun to proliferate in Finland during the late 1960s.

Perhaps most striking in the pamphlets was the specificity of concerns about different chemical substances in the air. This differed from complaints about smoke, dust and fumes that had been more or less regular in municipal public health boards in larger Finnish towns. It also differed from the concern over stuffy and stale air that had prevailed in nineteenth century hygienic thought. The hazard of urban air was now seen to lie in specific poisons such as carbon monoxide, hydrocarbons and lead. The pamphlets referred to the studies made by the FIOH and showed that in Helsinki the traffic alone produced 34,000 tons worth of carbon monoxide and 3,400 tons of hydrocarbons, while the quantity of lead in Finnish urban air had risen 800 per cent during the 1960s.[40] These substances were in turn feared for the specific health effects they had potential to cause. Lead and carcinogens were considered particularly worrisome, as they had the potential to cause long-term problems. Lead was known to accumulate in the environment and body. Hydrocarbons and other carcinogens were feared due to their potential to cause cancer and genetic mutations.[41] The Pollution Week protest in 1969 reproduced this same critical message by emphasising the quantities of specific pollutants and their hazards for health. In other words, the criticism about urban air quality was now more precise in terms of potential hazards and their causes than ever before.

The new critique also transcended the scale of the problem and laid it in a global context. The smog disasters in 1950s London, which had received little attention in Finland when they took place, were now presented as precedents and probable futures if nothing was done to remedy the situation. In these dystopian visions, people were forced to stay inside in ventilated spaces, use gas

39. Räsänen 2012, pp. 159–81.

40. Manula 1969, pp. 45–47.

41. Manula 1969, pp. 45–47; Valtiala 1969, pp. 19–45; Nordberg 1970, pp. 91–94.

masks or protect cities with ventilated domes.[42] The entire global atmosphere could even be doomed to be uninhabitable if the proliferation of cars and industrial production continued in populous developing countries.[43] More to the point, urban air pollution was not regarded as a separate public health issue. It was part of the anxiety about the overall destruction of the environment. Air pollution was one problem amongst a plethora of issues such as massive clear-cutting of forests, the pollution of coastal and inland waters, and drainage of marshlands.[44] Pollution was also connected to predictions about the rise of carbon dioxide in the atmosphere, which would eventually lead to warming of the global climate.[45] These issues were not merely simultaneous but also entangled, as they were seen to derive from the same pathological development in modern western society. The shift from vague concerns over smokes and fumes to chemical compounds in air also enabled the infusion of air pollution into wider concern over chemicalisation of the environment.

The new concern for air quality in Finland was dependent on local studies made by FIOH. These studies provided the numbers and facts that could be used to highlight local problems, but also to connect the state of local air with global developments. This did not mean, however, that the researchers were necessarily allies of the critics. The FIOH became a target of criticism despite its pioneering role in Finnish air pollution research. The institution's studies were heavily criticised while it itself was accused of protecting the interests of industry.[46] FIOH researchers argued that many of the concerns raised by the environmentalists were either unfounded or at least not acute in Finland. Often the same local studies were used by both parties to argue opposite points. While critics emphasised the sheer quantities of different toxic substances in air, FIOH researchers emphasised the toxicological principle that 'the dose makes a poison'. As the head of the institution argued, environment can handle reasonable pollution but not unreasonable.[47]

The different views were most evident in the case of lead poisoning. The health hazards of lead were well-known by the early twentieth century in occupational medicine, but, due to extensive studies indicating that it posed no danger in small doses, the use of tetraethyl lead in gasoline was approved

42. Manula 1969, p. 46; Valtiala 1969, p. 22.
43. Valtiala 1969, p. 20.
44. Ibid.; Pakkanen 1970.
45. Valtiala 1969, p. 32.
46. Launis 1972, p. 37.
47. Noro 1969, p. 126–30.

in the United States in the 1920s and its usage subsequently spread to other countries, including Finland.[48] In 1960, FIOH researchers stated in their report that there was no cause to believe tetraethyl lead in gasoline posed a danger to health.[49] The issue was invigorated again in the mid-1960s for various reasons. Geochemist Clair Patterson showed that the contamination of the environment from leaded fuel was considerably higher than had been previously estimated.[50] At the same time, the concern about lead contamination was given extra punch by the theory about the fall of the Roman Empire due to lead poisoning.[51] FIOH researchers attempted to mitigate these concerns and pointed to the weak evidence behind any claims about potential health effects. Critics saw the issue differently. Some regarded lead as a poison that slowly degenerated the human race through chronic diseases and genetic injuries. They also referred to American studies suggesting that increasing lead contamination could have an effect on precipitation and even alter the balance of the global climate.[52]

The attitudes of the FIOH researchers can be seen to depict the toxicological logic that Nash, for example, has criticised. It should be noted, however, that the new criticism was armed with the same knowledge and emphasis on quantities. By breaking urban air into numbers and chemical formulas, transnational air pollution research also provided the means for a new kind of critical rhetoric about urban air pollution in Finland, in which the quality of local air could be easily compared not only with other locales, but also with other aspects of the toxicity of the environment.

Air pollution as the new disease of civilisation

Uekötter has argued that fear about the health effects of air pollution in the late 1960s was a new phenomenon that differed, at least in the United States, from past attitudes and concerns.[53] It is easy to see similar novelty in the Finnish concern over urban air pollution in the late 1960s. The global scale of the concern, the fear of specific poisons and the alarmist rhetoric made the late 1960s a distinct period with regard to urban air pollution. On the other hand,

48. Warren 2019, pp. 105–20; Uekötter 2004, pp. 125–28.
49. Etylisoidun moottoripolttoaineen myrkyllisyys. Lausunto Esso Oy Ab:lle Helsingissä 26 Feb. 1960, Archives of the FIOH. Helsinki.
50. Warren 2019, pp. 115–26.
51. Gilfillan 1965, pp. 53–60.
52. Valtiala 1969, p. 32.
53. Uekötter 2009, p. 135.

the new criticism can also be seen to possess many continuities with older forms of concern over polluted air.

First of all, the new concern for air pollution can be seen as a continuation of the reductionist bacteriological view on health rather than adopting a more holistic point of view. As has been argued, the essential development in turn-of-the -century medicine was not that microbes cause diseases, but that diseases are entities defined by their causes.[54] This etiological point of view was also essential in industrial hygiene and toxicology as these defined diseases caused by specific chemicals.[55] It could be argued, then, that the late 1960s critics of air pollution were similar to the late nineteenth century hygienic movement when it embraced germ theory in order to specify its enemies from a plethora of vague notions about a dirty environment.[56] In the same way the 1960s health-conscious urban middle class embraced the view from industrial toxicology that could specify non-microbial concerns about the modern environment.

The result was also much the same. The fact that bacteria are everywhere meant that the hygiene movement had to focus on general cleanliness, despite having specified its enemies.[57] Similarly, the ubiquitous nature of toxic substances in a modern society was the fundamental enemy of the late 1960s critics. An essential feature of this rhetoric was that the modern world was thoroughly poisoned, as traces of potentially dangerous chemical components could be found almost anywhere – from atmosphere to foodstuffs. Although many experts on air pollution emphasised the need for mankind to adapt to the inevitable toxicity of modern world, critics called for action to turn the direction of society.

In other words, the late 1960s concern about urban air pollution had at its heart the same idea that had been at the core of nineteenth century concern about stale air in cities – namely the idea that progress and civilisation cause detrimental effects for humanity. As Rosenberg has argued, the anxiety over progress that had, in the nineteenth century, focused on cities was transferred into global and ecological thinking in the late twentieth century.[58] It should be added that what was essential in this transfer was the toxicological analysis of the environment. It was the nearly ubiquitous threat of toxins in the modern world that fuelled the concern over urban air quality in Helsinki, as well as

54. Carter 2017.

55. Sellers 1994, pp. 55–83.

56. Latour 1993.

57. Ibid.

58. Rosenberg 1998, p. 726.

Rachel Carson's concern about pesticides. As the Finnish pamphlets indicate, despite the global scale of the anxiety in the late 1960s, the modern city had not lost its place as a locus of anxiety even as civilisations' problems multiplied. In other words, rather than fundamentally changing, the concern over urban air in Finland had taken a new form whereby the intuitive concern over stuffy and dirty environments was replaced with a no less intuitive fear of toxins in the air.

Bibliography

Archives

Archives of the Finnish Institute of Occupational Health

Helsinki City Archives

Literature

Ahlberg, Eetu. 2019. 'From livelihood to a hazardous waste – a case study of asbestos mine in Paakkila'. Master's Thesis, Department of Geographical and Historical Studies, University of Eastern Finland.

Carter, Kay Codel. 2003. *The Rise of Causal Concepts of Disease: Case Histories*. Farnham: Ashgate.

Gilfillan, Colum. 1965. 'Lead poisoning and the fall of Rome'. *Journal of Occupational Medicine* 7 (2).

Harjula, Minna. 2003. *Tehdaskaupungin takapihat. Ympäristö ja terveys Tamperella 1880–1939.* Tampereen historiallisen seuran julkaisuja XVII, Tampere.

Heiman, Harry. 1967. 'Status of air pollution health research, 1966'. *Archives of Environmental Health* **14**

Homburg, Ernst and Elisabeth Vaupel. 2019. 'A conceptual and regulatory overview, 1800–2000'. In Ernst Homburg and Elisabeth Vaupel (eds). *Hazardous Chemicals: Agents of Risk and Change, 1800–2000.* New York: Berghahn Books.

Hosiaisluoma, Väinö. 2001. 'Jäkälät, Sammalet, puut ja perhoset ilmanlaadun ilmaisijoina'. In Simo Laakkonen, Sari Laurila, Pekka Kansanen and Harry Schuman (eds). *Näkökulmia Helsingin ympäristöhistoriaan.* Helsinki: Edita.

Hueper, Wilhelm. 1952. 'Environmental cancer hazards caused by industrial air pollution'. In Louis McCabe (ed.) *Air Pollution. Proceedings of the United States Technical Conference on Air Pollution.* New York: McGraw-Hill Book Company.

Jas, Nathalie. 2015. 'Adapting to reality'. In Nathalie Jas and Soraya Boudia (eds). *Toxicants, Health and Regulation Since 1945.* New York: Routledge.

Jauho, Mikko. 2007. *Kansanterveysongelmien synty. Tuberkuloosi ja terveyden hallinta Suomessa ennen toista maailmansotaa. Helsinki: Tutkijaliitto.*

Jormalainen, Aulis, Arvo Laamanen and P.U. Lehtinen. 1961. 'An investigation of air pollution in the city of Helsinki in 1959'. *Suomen kemistilehti* A **34** (9): 152–64.

Jütte, Robert. 2018. 'Reodorizing the Modern Age'. In Mark Smith (ed.) *Smell and History: A Reader.* Morgantown: West Virginia University Press.

Kajanne, Paavo and Stiven Laiho. 1958. 'A preliminary investigation of air pollution in Helsinki with particular attention to diesel smoke'. *Suomen kemistilehti* B **31** (4): 193–98.

Kehoe, Robert. 1952. 'Effects of prolonged exposure to air pollutants'. In Louis McCabe (ed.) *Air Pollution. Proceedings of the United States Technical Conference on Air Pollution.* New York: McGraw-Hill Book Company.

Kirchhelle, Claas. 2018. 'Toxic tales – recent histories of pollution, poisoning, and pesticides (ca. 1800–2010)'. Essay review. *NTM Zeitschrift für Geschichte der Wissenschaften, Technik und Medizin* 26: 213–29.

Kolbe, Laura. 2007. 'Matkalla sivityneistöön'. In Katriina Järvinen and Laura Kolbe (eds). *Luokkaretkellä hyvinvointiyhteiskunnassa: nykysukupolven kokemuksia tasa-arvosta*. Helsinki: Kirjapaja.

Kruut Marja. 1999. 'Savua ja Nokea'. In Simo Laakkonen, Sari Laurila, Marjatta Rahikainen and Päivikki Kallio (eds). *Nokea ja Pilvenhattaroita. Helsinkiläinen ympäristö 1900-luvun vaihteessa*. Helsinki: Helsingin kaupunginmuseo.

Laamanen, Arvo and Leo Noro. 1967. *Helsingin ilman saasteen lähdeyhtymäkatsaus ja siihen liittyvät kaupunkikohtaiset ilmansuojelunäkymät*. Työterveyslaitoksen tutkimuksia n:o 34. Helsinki: Työterveyslaitos.

Lahti, Vesa and Ilkka Saarela. 1991. 'Kun vesi on myrkkyä. Tapaustutkimus myrkkyonnettomuudesta Kärkölän Järvelässä'. In Ilpo Massa and Rauno Sairinen (eds). *Ympäristökysymys. Ympäristöuhkien haaste yhteiskunnalle*. Helsinki: Gaudeamus.

Lahtinen, Rauno and Timo Vuorisalo. 2004. '"It's war and everyone can do as they please!" An environmental history of a Finnish city in wartime'. *Environmental History* **9** (4): 659–700.

Latour, Bruno. 1993. *Pasteurization of France*. Trans. by Alan Sheridan. Cambridge: Harvard University Press.

Launis, Tapani. 1971. *Helsinki: Yleiskaavaehdotus 1972. Ympäristönsuojelun esiselvitykset Helsingin 1972–1973 yleiskaavoituksessa*. Helsinki: Kaupunkisuunnitteluvirasto, yleiskaavaosasto.

Lehtonen, Turo-Kimmo. 1995. *Puhtaan elämän jäljillä. Huoli hygieniasta suomalaisissa terveydenhoitolehdissä 1889-1900*. Helsinki: Kuluttajatutkimuskeskus.

Nylander, M.W. 1866. 'Les Lichens Du Jardin Du Luxembourg'. *Bulletin de la Société Botanique de France* **13** (7) 364–71.

Mäkiranta, Janne. 2022. *Clarifying the Air: Finnish Air Pollution Experts and the International Quest for Safe Air, 1940s-1970s*. Turku: Annales Universitatis Turkuensis.

Maunula, Leena. 1969. 'Yksityisautolla joukkohautaan'. In Leena Maunula (ed.) *Alas auton pakkovalta*. Helsinki: Tammi.

Nash, Linda Lorraine. 2008. 'Purity and danger: Historical reflections on the regulation of environmental pollutants'. *Environmental History* **13** (4): 651–58.

Nordberg, Rainer. 1969. 'Auto – Tappava leikkikalu'. In Jukka Pakkanen (ed.) *Minne kukat kadonneet*. Helsinki: Tammi.

Noro, Leo and Arvo Laamanen. 1958. 'Ulkoilman saastumisesta Suomessa'. *Duodecim* 4.

Noro, Leo. 1969. 'Ympäristön pilaantuminen ja terveys'. *Sosiaalilääketieteellinen aikakauslehti* 7 (125).

Phair, John J. 1956. 'The epidemiology of air pollution'. In Paul Magill, Francis Holden and Charles Ackley (eds). *Air Pollution Handbook*. New York: McGraw-Hill Book Company.

Räsänen, Tuomas. 2012. 'Converging environmental knowledge: Re-evaluating the birth of modern environmentalism in Finland'. *Environment and History* **18** (2): 159–81.

Rosenberg, Charles E. 1998. 'Pathologies of progress: The idea of civilization as risk'. *Bulletin of the History of Medicine* 72 (4): 714–30.

Saarikangas, Kirsi. 1998. 'Suomalaisen kodin likaiset paikat: hygienia ja modernin asunnon muotoutuminen'. *Tiede & Edistys: monitieteinen aikakauslehti* **23** (3): 198–220.

Schönach, Paula. 2008. *Kaupungin savut ja käryt: Helsingin ilmansuojelu 1945–1982*. Helsinki: Helsingin yliopisto.

Sellers, Christopher. 1994. 'Factory as environment: Industrial hygiene, professional collaboration and the modern sciences of pollution'. *Environmental History Review* **18** (1): 55–83.

Sellers, Christopher. 1997. *Hazards of the Job: From Industrial Disease to Environmental Health Science*. Chapel Hill & London: University of North Carolina Press.

Sellers, Christopher. 2012. *Crabgrass Crucible: Suburban Nature and the Rise of Environmentalism in Twentieth Century America*. Chapel Hill: University of North Carolina Press.

Thorsheim, Peter. 2006. *Inventing Pollution: Coal, Smoke, and Culture in Britain since 1800*. Athens: Ohio University Press.

Uekötter, Frank. 2004. 'The merits of the precautionary principle: Controlling automobile exhausts in Germany and the United States before 1945'. In E. Melanie DuPuis (ed.) *Smoke and Mirrors: The Politics and Culture of Air Pollution*. New York: New York University Press.

Uekötter, Frank. 2009. *The Age of Smoke Environmental Policy in Germany and the United States, 1880–1970*. Pittsburgh: University of Pittsburgh Press.

Valtiala, Nalle. 1969. *Varokaa ihmistä. Orig. Varning för människan*. Trans. Helena ja Martti Linkola. Porvoo: WSOY.

Warren, Christian. 2019. 'Old situations, new complications'. In Ernst Homburg and Elisabeth Vaupel (eds). *Hazardous Chemicals: Agents of Risk and Change, 1800–2000*. New York: Berghahn Books.

Chapter 11

FROM ERADICATION CAMPAIGNS TO 'CARE PROTECTION': FINNISH ENDANGERED ANIMALS IN THE TWENTIETH CENTURY

Tuomas Räsänen

Introduction

Finns often pride themselves in living close to nature and thus cherishing their living environment. There is definitely some truth in the first half of this claim, while the scale of endangered habitats proves that the latter claim is necessarily not true.[1] The country was a latecomer in urbanisation when compared to early industrialised Western Europe. Until the latter part of the twentieth century, a large proportion of the populace continued to live in the countryside, while drawing their livelihood directly from the land and Finland's abundant watercourses. Since the urbanisation process intensified after the Second World War, newly urbanised people have remarkably kept their ties to the land. Today, according to Statistics Finland, there are more the 500,000 summer cottages in a country of circa 5.5 million inhabitants,[2] most of which are situated deep in the countryside, preferably on the shores of a lake or the sea. Among European countries, only Norway compares to Finland in this sense.[3]

1. For endangered habitats, see ympäristö.fi, joint webpages of the Finnish environmental administration, Uhanalaistuminen jatkuu lähes kaikissa elinympäristöissä: https://www.ymparisto.fi/fi-FI/Kartat_ja_tilastot/Ympariston_tilan_indikaattorit/Luonnon_monimuotoisuus/Uhanalaistuminen_jatkuu_lahes_kaikissa_e(61090) (accessed 21 Sept. 2021)

2. Statistics Finland, Freetime Residences 2020: https://www.stat.fi/til/rakke/2020/rakke_2020_2021-05-27_kat_001_en.html (accessed 21 Sept. 2021)

3. For Norwegian summer cottage culture and its junctures with environmental questions, see Gansmo et al. (eds) 2011.

doi: 10.3197/63824846758018.ch11

It is, therefore, no wonder that the Finnish identity-building has been very much tied to ideas about the environment. On one hand, there is a reminiscence of a pioneer mentality, in which the environment is something to be conquered and the others who inhabit the land – that is, animals – are seen as either resources or risks. On the other hand, the environment has increasingly been seen as the source of country's success and virtues, and certain animal species embody the best qualities that the Finns see in themselves. This ambivalence towards wild animals has particularly characterised the history of twentieth century Finland, when the old and new mentalities have co-existed, often causing conflicts about the use of the environment.

This chapter focuses on the human relation to wild animals in Finland from the early twentieth century to the present day. Clearly, even in a small country with relatively low biodiversity, it would be a mission impossible in such a short space to comprehensively examine the history of wild animals spanning over one hundred years. Therefore, it is necessary to restrict the scope of the analysis. I will concentrate on dominant trends in the context of Finnish animal history and important turning points that represent evolutions of new ways relating to wild animals. By turning points, I do not mean to identify revolutionary events in the history of Finnish human-animal relations. As Frank Uekötter has reminded us, environmental historians are not doing justice to their objects of inquiry in seeking such events that separate 'before' and 'after' into totally different realities. He maintains that turning points are useful in periodising developments in environmental history but, instead of sharp ruptures, they are rather gradual shifts, where condensations of events sometimes accelerate the change.[4] Therefore, for those who lived through them, these developments could have gone almost unnoticed, but retrospectively they can be seen as changing the world. In this vein, I argue that there have been distinct, yet overlapping developments, in the twentieth century history of human relations to wild animals in Finland.

The chapter begins with the examination of developments from the late nineteenth century and the early twentieth century, which were characterised by the dichotomy between useful and harmful animals. This will be followed by the emergence of the second turning point and the growing concern over certain endangered species. The late twentieth century, when the third turning point took place, saw an unprecedented effort and new techniques to save endangered species. However, apart from conservationists (and lofty declarations), this awakened concern for species protection was still restricted to a

4. Uekötter 2010.

rather small number of species, while a vast majority of endangered species were left behind and seen as having value only as part of ecosystems, if at all. Each of these turning points will be highlighted by focusing on certain animal species that are representative of their groups. The large picture of the Finnish human-animal relations is mainly synthesised from the research literature. When discussing individual animals as examples of change and continuities, I have complemented the narrative with empirical analysis of influential contemporary commentaries about animals.

Conflict over resources

The subspecies of reindeer that inhabits Finland and northwestern Russia is appropriately named the Finnish forest reindeer (*Rangifer tarandus fennicus*). This species has had a special place in the history of human settlement in the area that we nowadays know as Finland. When present-day Finland was populated by human settlers after the Last Glacial Period some 12,000 years ago, they are said to have been following reindeer. For thousands of years, reindeer (and moose) were the most important prey for humans and, along with abundant fish, enabled humans to survive in harsh northern climes where agriculture could provide only partial and insecure subsistence.[5] The heavy hunting pressure had consequences. In highly populated southwestern Finland, reindeer and moose were probably already rare during the Middle Ages. During the following centuries these populations continued to shrink and, by the early twentieth century, the wild reindeer was hunted to extinction in Finland.[6]

One cannot escape a strong symbolism in the destruction of the Finnish forest reindeer population; the species that enabled the human population to survive and colonise the land area became the victim of overhunting. The reindeer was not the only species to face this destiny. Several species that were valued by humans, either for food or, in the case of fur animals such as Eurasian beaver, Eurasian otter and stoat, for commercial value in European markets, were all decimated by excessive hunting. The beaver population was wiped out entirely by the late nineteenth century, while the rest scraped by in the sparsely populated areas of Eastern and Northern Finland.[7]

---

5. Huurre 1998, pp. 153–82.
6. Metsähallitus, Metsäpeura: https://www.suomenpeura.fi/fi/metsapeura/levinneisyys.html; Tourunen 2008, pp. 114–15.
7. Rantaniemi 1901, pp. 36, 50.

It has been argued that the human craving to dichotomise animals into two opposite categories dates back to the prehistoric times and is probably written in human biology. At one end of the spectrum, there are those animals that are seen as important resources. The other end consists of enemies, noxious animals that threaten human life and livelihoods.[8] (Between these extremes, there are, of course, a vast number of species that were met with indifference, either because they were from the human point of view useless or because they were invisible). Sometimes, though, these categories overlapped. Seals, for example, fell neatly into both categories, as they were hunted for their meat and train oil by coastal communities but, as predators, they were (and are, by some) also hated for preying on fish.[9]

When the central government tightened its grip on the country's natural resources in the late nineteenth century, no room was left for animals that were useless or, even worse, harmful. By going after the same prey as humans, or just by being intimidating or otherwise objectionable, these animals were direct threats to civilisation. They were nature's mistakes, that must be corrected by getting rid of them. There was little new in persecuting offending animals. The government and municipalities had paid bounties for wolves and bears since the Middle Ages and obliged locals to participate in organised hunts whenever needed. Gradually the bounty system was extended to all predatory animals, as well as to species that were considered somehow objectionable, and persecution was normalised as a systematic and incessant campaign, propagated and financed by the administration. By the late nineteenth century, eradication of unwanted animal species had become one of the cornerstones of building a modern and prosperous state. The list of animals to be eradicated consists of dozens of species, from large predators and every kind of predatory bird even to tiny seed-eating birds, such as sparrow.[10] It did not matter if the species was already rare and thus of little harm to human economies.

Soon many of them were indeed rare. In the late nineteenth century, wolves still roamed in hundreds all over Finland. Since cervids, their important prey, were all but gone, wolves often gravitated near human dwellings with tragic consequences. Even in present-day Finland, it is often recalled how wolves, in fact probably only one individual wolf, killed more than twenty children in southwestern Finland in 1880–1882.[11] After a few decades of obses-

8. Vuorisalo and Oksanen 2020.

9. Ylimaunu 2000, pp. 109–27.

10. Ilvesviita 2005, pp. 144–53. Mykrä et al. 2005.

11. Teperi 1977. Lähdesmäki 2015, p. 187.

sive slaughtering of wolves, except for lone wandering specimens, wolves were forced to retreat into the sparsely inhabited eastern and northern parts of the land.[12] Ditto the other large predators – brown bear, wolverine and lynx. The numbers of animals killed were staggering. Hunting statistics reveal that, in the 1910s, the most hectic phase of the 'War Against Seals', as it was dubbed by fishery scientists, approximately 10,000 marine seals, grey seals and Baltic ringed seals, were killed annually in Finland alone. In the Northern Baltic Sea area, mainly in Finland, the Soviet Union and Sweden, humans killed more than 600,000 marine seals in total throughout the twentieth century, of which by far the biggest toll was taken in the first half of the century.[13] At the beginning of the twentieth century, there were possibly more than 300,000 seals in the Baltic Sea. By the 1970s, the number had sunk to less than 5,000.[14] These numbers are very comparable to the much better-known overkills of, for example, American Bison.

Turning the tide

The war against seals continued well into the latter part of the twentieth century, as did the eradication campaigns against some other predators. Finland has over 1,000 kilometres of land border with Russia/Soviet Union, which mostly runs through the sparsely inhabited boreal forests without any natural barriers. The Soviet Union provided Finland throughout the twentieth century with an ample reservoir of animal intruders, some of which were greeted mostly with warmth, such as the Finnish forest reindeer, which started to slowly reappearing on the Finnish side of the border from the 1940s, while others were not. The wolf population in Finland was kept alive only by the constant flux of migrating individuals from the vast forests of Soviet Karelia. When spotted in Finland, they all got shot sooner or later.[15] There is nothing new under the sun, in this regard.

The long twentieth century, with regard to animal history, started with the urge to modernise the country. The crucial events were the hunting laws of 1868 and 1898, that codified and institutionalised government-led campaigns

12. Lähdesmäki 2020. Finland was not by any means the only country to eradicate its wolves, for similar campaigns were launched all over the industrialising world. For the campaign to eradicate wolves from the United States, for example, see Robinson 2005.
13. Gottberg 1921; Härkönen et al. 1998; Martti Soikkeli, N.d, Memorandum, Antti Halkka's Collections, WWF Finland Working Group on Marine Seals.
14. Härkönen et al. 1998.
15. Bisi 2010; Lähdesmäki 2020.

to eradicate all harmful species.[16] This advent of institutionalised killing of animals was the first turning point in the history of human relations to wildlife. The second turning point emerged during the period from the 1920s to the 1960s with a new interest in protecting some endangered species. During this period, some species were gradually freed from the stigma of being nature's greatest mistakes, while others were elevated from mere resources to having a right to be respected dwellers in Finnish nature.

In the early years of the century, some Finnish biologists, following the international discussion on the matter, had already suggested that predators were not villains that terrorised the harmony of nature and threatened at every moment to throw civilisation into chaos, as they were presented in popular imagination. These new ideas were put in practice with the country's first law on natural protection, enacted in 1923, which omitted many predatory birds from the list of exterminated species and instead gave them a protected status. The reason behind this volte-face was that these species were understood to be of minor or no threat to humans and their domestic animals. For example, all predatory birds had been listed as noxious animals just because some species preyed on important game birds. Moreover, many of them were so rare that the discourse of threat had lost its argumentative power.[17] But it would be oversimplifying to define this law as an exact and decisive turning point in animal history. Among the scientific community, on whose authority the law rested, the change had already been unfolding for some time. Even more importantly, the law did not do much to change attitudes among local people, and often hatred towards and killing of protected predators continued unabated for several decades.[18]

The early ideas of natural conservation in Finland, as elsewhere in Europe, found their justification in history and cultural heritage (along with scientific values).[19] It is no wonder, then, that the species that were first seen as worthy of protection were those with cultural and historical meaning for the Finns. Not only conservationists but even some hunters advocated the protection of the brown bear, a sacred species in Finno-Ugric religions, from the early twentieth century. This was in striking contrast with the treatment of wolf, another predator that had been ferociously hunted for centuries, which was

16. Ilvesviita 2005, pp. 148–49.

17. Ibid., pp. 230–33; Pohja-Mykrä 2014, pp. 34–35.

18. See Räsänen 2020, p. 29.

19. Pekurinen 1997.

unanimously despised by everyone, including conservationists, well into the latter part of the twentieth century.[20]

The dispute over the bear indicated how difficult it was to dislodge prevailing attitudes towards predators, no matter how rare they actually were. It was easier to accept or even embrace the protection of animals that posed danger to no-one, especially if symbolic and identity-related meanings coud be attached to them. The case in point in this regard was the whooper swan (*Cygnus cygnus*). Immortalised in the first known rock art sites and referred to in countless folk tales, starting with the Finnish national epic Kalevala where it guarded the river separating our world from the underworld, the Whooper swan has a special place among birds in Finnish culture. In Finno-Ugric mythology, the swan was related to humans and thus sacred. Despite this intimate relationship, the swans were hunted and eaten for subsistence (though not among Finno-Ugric tribes in Russian Karelia, where the mythological relationship endured longer) and their eggs were stolen and sold to collectors. By the mid-twentieth century, the swan population was decimated to a mere fifteen pairs. After relentless hunting pressure, swans had learned to fear humans. This was why all swans in Finland nested in the far north of Lapland behind the roadless stretches of fells and swamps, as far from humans as possible.[21] There is a tragic irony involved in this swan behaviour. While wintering in Denmark and northern Germany, they could be spotted right next to human occupation without showing any particular fear of humans. It turned out that they did not escape people *per se* but only Finnish people, in the land occupied by the very group of people that used to totemise the bird. Obviously they had learnt to fear humans during their nesting season in Finland. Because of this alleged wildness, Finns imagined that swans truly preferred the solitude and peace of remote wilderness. As with many other so-called wilderness animals, nowadays it is clear that they do not mind living next to people, and the population is strongest in the most inhabited – and most fertile – part of the country, as long as they do not have to fear death.

Despite official protection of the Whooper swan in 1934, illegal hunting and gathering of eggs continued. It did not help either that, during the Continuation War, Finnish soldiers sometimes shot swans for food in Russian Karelia, where the population was somewhat stronger. Then, in the 1950s, it came about that the demise of the swans, which in the 1930s and 1940s had concerned only a handful of conservationists, became a widely discussed

20. Lähdesmäki 2020, pp. 261–62; Kalliola 1958, p. 429.

21. Merikallio 1950.

topic, even among the general public, and swans were adopted as an animal darling for thousands of Finns. The astonishing rise of swan in Finland has in many instances been attributed to the work of Yrjö Kokko, who worked as a veterinarian in the Lappish parish of Muonio, when he set off to find and photograph the last of the swans. After years of tiresome searching, he and his companion, a local man, finally managed to find a pair by an unnamed wilderness lake (Kokko never told publicly the name or the place of the lake so that hunters or egg-thieves could not find the pair). After returning home he wrote a book, a true love story, about his encounters with the swan pair, 'Hanna' and 'Marski'.[22]

The book, *Laulujoutsen – Ultima Thulen lintu* (The Whooper Swan – the Bird of the Ultima Thule), came out at the moment, in 1950, when Finland was beginning to develop into a modern society and attitudes towards wild animals had been evolving for some time. Together with the sequel titled *Ne tulevat takaisin* (They Are Coming Back), published in 1954, the book became a bestseller, and the books' call for the protection of this bird so full of symbolism in Finnish culture stroke a chord with the public at large. In the following years, ordinary people were lining up en masse to see and admire, for example, a swan named 'Aino', a lone indvidual who tried to survive in Central Finland through the winter, and 'Hannu', which was released from the veterinary clinic into the river crossing the city of Joensuu in Eastern Finland. Only a few years earlier, it had not been uncommon for people to bait swans, when they happened to stop by, while migrating from or to nesting places in the North. Several veterinary clinic were established, by cities as well as by ordinary citizens, in different parts of Finland, where swans found during the winter could be taken care of.[23] The nation clearly had warmed to the Whooper swan and, by the turn of the 1960s, conservationists could rejoice in how the population had started to climb, though first only in Lapland and in the east near the Russian border.[24] The success of saving the whooper swan became one of the founding stories of the Finnish modern environmentalism. The history book of the Finnish Association for Nature Conservation, which is the biggest and the oldest environmental advocacy group in Finland, is aptly named as the 'The legacy of the whooper swan'.[25] Nowadays whooper swans are so common

22. Kokko 1950.

23. Eg. Ketola 1945; Erkamo 1955; 'Toimintaa joutsenen suojelemiseksi', *Suomen Luonto* 18 (1959): 88–89.

24. Eg. Haapanen and Helminen 1965; Haapanen and Helminen 1966.

25. Telkänranta (ed.) 2008.

that occasionally some commentators propose hunting them, not for meat but to restore the balance among different birds. These proposals have so far always been met with outrage and nearly universal condemnation.

Labouring for animals

The period from the early twentieth century to the 1960s can be seen as an incubation period[26] for the new relationship between humans and wild animals in Finland. For centuries, the relationship was characterised by the sharp distinction of beneficial and harmful animals. As the twentieth century went on, this dichotomy began to wane and was gradually replaced by a much more complex relationship. This did not mean that there was no longer hatred and enmity towards certain animal species, such as large carnivores. An increasing number of people, however, started to appreciate even the most dangerous of animals, arguing that, despite the fear they engendered, they deserved a right to exist in the Finnish nature alongside humans. Indeed, in this emerging mentality, many admired these animals precisely because they represented the wildness and the danger that made nature the opposite of overly civilised everyday life in an urbanised environment.

The traditional relationship between humans and wild animals had arisen from the notion of progress. The success of the nation rested on the effective use of its natural resources, animals included, and those animals who impeded achieving progress were destined to go. The emerging appreciative relationship, in contrast, emphasised co-existence and extension of the ethical sphere to the animal kingdom.[27] This change in human-wild animal relations is perhaps best exemplified by the use and reception of labour: who worked for whom. For thousands of years, the value of animals depended on the labour they did for humans. The most respected animals were those who provided shelter for humans and accompanied them in hunting, such as dogs; who provided support in agriculture and transport with their muscles and manure, such as horses and oxen; or kept pests in check, such as cats. In short, they did work for humans, and at the end of their lives they often provided humans with food, clothes and other utensils. In this sense, there has been an almost complete about-face in human-animal relations. Today, conservationists, government officials and ordinary citizens spend countless hours working to ensure the wellbeing of

---

26. The term incubation period has been borrowed from the Austrian cultural historian Egon Friedell and his treatise on European cultural history, Friedell 1927–31.

27. For the development of animal ethics, see, for example, Nash 1989.

wild animals. Remarkably, these animals, for which humans work, are often the same animals that we tried just a short while ago to get rid of. Instead of being harmful, though according to some they still are, more often now these species are characterised as charismatic animals.

As said earlier, turning points in environmental history rarely take place as sudden explosions but more often evolve gradually from tiny trickles before emerging as mighty rivers. Retrospectively, it is easy to see these trickles or the seeds of a new relationship emerging even back when the traditional way dichotomising of animals was the order of the day. Bird boxes, for example, had been a popular method in Finland and elsewhere of helping songbirds from the late nineteenth century.[28] Hunters toiled hard bringing extra food for game animals and improving their habitats.[29] Obviously, neither practice was directed at protecting endangered species. Bird lovers wanted to help cheering songbirds, who also ate damaging insects, but who were increasingly lacking nesting trees.[30] Hunters were motivated mainly by being able to continue killing animals.

When it comes to institutionalised protection of endangered species in Finland, an important milestone was the founding of World Wildlife Fund Finland in 1971. In 1972 WWF Finland launched conservation projects to protect five key species: the Finnish forest reindeer, the Saimaa ringed seal, the grey wolf, the peregrine falcon and the white-tailed eagle, all of which were iconic species to conservationists. The project of protecting the white-tailed eagle in particular was groundbreaking for its intensity and for innovative new conservation techniques, that perfectly illustrate the new relationship to wild animals.

By the early 1970s, the once ubiquitous white-tailed eagle (*Haliaeetus albicilla*) was inevitably heading towards extinction in the Baltic Sea area. Due to decades of persecution and human disturbance, the Finnish population of eagles had shrunk to a mere few dozen breeding pairs, which lived in three separate enclaves. While the persecution gradually eased, toxic chemicals, such as DDT, PCBs and methyl mercury, seemed to seal the eagle's destiny. Due to these chemicals, eagles had become unable to produce healthy offspring. Conservationists reckoned that soon there would be only old eagle individu-

28. Vuorisalo et al. 1999, pp. 116–18; Lehikoinen et al. 2020, pp. 436–39.

29. See for example, Moilanen and Vikberg (eds) 1986, pp. 15–27, 37–38.

30. Lehikoinen et al. 2020, pp. 436–38.

als left. When they eventually died, white-tailed eagles would roam across the archipelagic sky only in memories.[31]

Together with their Swedish colleagues at the Swedish Society for Nature Conservation, WWF Finland launched an extensive and multi-faceted project to save the eagles from extinction. They quickly realised that the traditional way of conserving species – that is, the preservation of their habitat from hunting and disturbances – would not be enough to save the eagles. Even if left in peace, eagles would still eat toxic fish and waterfowl. Therefore, conservationists had to invent novel conservation methods, including feeding the eagles with clean food (mainly pig carcasses), building them artificial nests in tranquil places and nurturing wounded eagle individuals.[32] Similarly to the protection of the whooper swan, the white-tailed eagle project has been a real success story. The species has been removed from the list of endangered species and spotting individuals has become commonplace.

The white-tailed eagle project was not the first time that humans had helped animals, for example, by feeding them. But this was, to my knowledge, the first time in Finland and perhaps even anywhere that humans had actively and intensively tried to assist the survival of endangered wild animal populations by taking intensive care of individual animals. Hence, I have elsewhere termed this wholly new method of conservation 'care protection'.[33] Remarkably, in this new conservation ethos, the labour of care was done for species, such as the white-tailed eagle, that only recently had been objects of hate and killing.

Subsequently, care protection has been introduced as a method for conserving several other animal species, in Finland and elsewhere, whenever traditional protection has proved insufficient. A good example in Finland is the Saimaa ringed seal (*Pusa hispida saimensis*), the subspecies that lives only in the Finnish lake of Saimaa, which, similarly to its marine cousins, depends on ice and snow for survival. In recent years, however, too often snow and ice have been scarce or even missing, which endangers the reproduction of the seals and the survival of the whole population. Therefore, in mild winters conservationists have, with great effort, heaped up snowdrifts for seals to build their lairs and also constructed artificial lairs for seals.[34] Ringed seals, namely the subspecies Baltic ringed seal (*Pusa hispida botnica*), are also troubled by the

31. Koivusaari et al. 1973.
32. Stjernberg 1995; Wallgren 2016, pp. 13–14.
33. Räsänen, 2020.
34. Jaakkola et al. 2018, pp. 140–41.

changing climate in the Baltic Sea.[35] Interestingly, however, no such efforts to care and labour on their behalf have been made to help marine seals. This demonstrates that concern for, let alone intensive care of, animals is very unevenly and haphazardly distributed. Even animals that are almost identical, such as lake seals and marine seals, are treated differently depending on the meanings humans attach to them. The Saimaa ringed seal lives solely in Finland, which makes it an important symbol of the Finnish nature, while the Baltic ringed seal is just one animal species among many.

Conclusions – new dichotomies

The history of human relations to wild animals in Finland is possible to construe as a story of increasing tolerance of animal lives and extension of the species that are considered worthy of protection and care. This would, however, oversimplify the past. Eradication campaigns are long gone and bounties for dead animals are no longer paid, but this has not silenced the debate over, for example, whether two or three hundred wolves are too many, while illegal hunting has remained widespread.[36] Many of the formerly overhunted species are today protected, but the protection merely aims at achieving a so-called favourable conservation status for these species, which in effect is an euphemism for keeping their numbers at the minimum, defined by scientists, without risking extinction. Although dozens of protected areas have been established during the course of the twentieth and twenty-first centuries, habitats continue to decline due to land-use practices – and increasingly because of climate change. For these reasons combined, currently one in nine species in Finland is endangered, as assessed by Finnish Ministry of the Environment and Finnish Environment Institute.[37]

Conservationists have every reason to celebrate the success of eagles and other previously endangered species that have bounced back in recent decades. The old dichotomies, where animal species were categorised into useful and harmful species have waned, though not entirely disappeared. Yet, old dichotomies have been replaced by new ones. There seems to be an almost unanimous consent that invasive species are the present-day villains, which are free for all to kill. But there are also more subtle dichotomies, that are related somewhat loosely to the charisma of animal species. The recovery of animal populations has overwhelmingly encompassed certain species, which in social construction

35. Markus Meier et al. 2004.

36. Pohja-Mykrä and Kurki 2014.

37. Hyvärinen et al. (eds) 2019, p. 25.

have been deemed as having more worth than others. Therefore, they also tend to attract most attention from conservationists, politicians and the wider public alike, while countless other species have been more or less ignored, regardless of their conservation status.

However, it is far from evident what makes species charismatic, as there seems not to be a single denominator. Large size and cuteness definitely help. Predators may be more often recognised as charismatic species than herbivores. Some of the reasons are culturally specific and are related to concepts of representation and identity, as the case of the whooper swan demonstrates. Then, there are charismatic species that fall into none of the above-mentioned categories. One such is the river pearl mussel (*Margaritifera margaritifera*). A species that remains virtually all its life stuck at the river bottom is a highly unlikely candidate for the group of charismatic animals, but it has nonetheless been recently adopted as a darling species of conservationists.[38] The fact that it is the oldest living animal species in Finland, the pearl it sometimes carries and its dependence on salmon, arguably the most charismatic of all fish species in Finland, may partly explain its worth. This list of denominators is, of course, far from conclusive, and calls for further research examining cultural appreciations of animals. Rather than being systematic analysis, these examples merely intend to show how unbalanced and coincidental attitudes towards different species are.[39]

This chapter began with the widespread claim among the Finnish to be a particularly nature loving nation. The existence of these new dichotomies and the ongoing decline of animal populations makes these claims rather dubious. To give just one example, each of the Baltic states has a wolf population roughly the same size or only slightly smaller than that of Finland, despite being more than five times smaller in land area.[40]

We are living amidst the sixth mass extinction of our planet's history. If we are to keep the planet habitable for humans and other living beings alike, we need to care not only for species that are randomly chosen as culturally valuable but also those that are not charismatic or even visible. The history of human relations to wild animals in Finland has experienced three major turn-

38. See for example, WWF, 'Jokihelmisimpukka': https://wwf.fi/elainlajit/jokihelmisimpukka/ (accessed 4 Nov. 2022).

39. For more detailed discussion on charismatic species in conservation biology, see Albert et al. 2018.

40. International Wolf Center, Wolves of the World: https://wolf.org/wow/world/ (accessed 11 Nov. 2022).

ing points during the long twentieth century, which all conferred on a growing number of species the right to exist unmolested. Another turning point is required, one that will discard the remaining dichotomies and recognise without cultural prejudices and privileges the value of every animal species for the whole planet. Considering the discussion about biodiversity loss, which has intensified in recent years, we may well be witnessing such a new change in human-animal relations.

Bibliography

Albert, Celine, Gloria M. Luque and Franck Courchamp. 2018. 'The twenty most charismatic species' *PLoS ONE* **13** (7): https://doi.org/10.1371/journal.pone.0199149

Bisi, Jukka. 2010. *Suomalaisen susikonfliktin anatomia.* Oulu: Oulun yliopisto.

Erkamo, V. 1955. 'Ne tulevat takaisin', book review. *Suomen Luonto* 14: 116–117.

Friedell, Egon. 1927–31. *Kulturgeschichte der Neuzeit. Die Krisis der europäischen Seele von der schwarzen Pest bis zum Weltkrieg.* München: Beck.

Gansmo, Helen Jøsok, Thomas Berker and Finn Arne Jørgensen (eds). 2011. *Norske hitter i endring.* Trondheim: Tapir Akademisk forlag.

Gottberg, Gunnar. 1921. 'Sälfånsten år 1909–1918'. *Fiskeritidskrift för Finland* 28: 16–18.

Haapanen, Anti and Matti Helminen. 1965. 'Joutsenen nykylevinneisyys'. *Suomen Luonto* 24: 39–42.

Haapanen, Anti and Matti Helminen. 1966. 'Joutsentietoja vuodelta 1965'. *Suomen Luonto* 25: 65–66.

Härkönen, Tero, Olavi Stenman, Mart Jüssi, Ivar Jüssi, Roustam Sagitov and Michael Verevkin. 1998. 'Population size and distribution of the Baltic Ringed Seal (*Phoca hispida botnica*)'. *NAMMCO Scientific Publications* 1.

Huurre, Matti. 1998. *Kivikauden Suomi.* Helsinki: Otava.

Hyvärinen, Esko, Aino Juslén, Eija Kemppainen, Annika Uddström and Ulla-Maija Liukko (eds). 2019. *The 2019 Red List of Finnish Species.* Helsinki: Ympäristöministeriö and Suomen ympäristökeskus.

Ilvesviita, Pirjo. 2005. *Paaluraudoista kotkansuojeluun: Suomalainen metsästyspolitiikka 1865–1993.* Rovaniemi: Lapin Yliopisto.

Jaakkola, Minttu, Timo Vuorisalo and Lasse Peltonen. 2018. *Saimaannorppa ja ihminen: Sosiaalisesti kestävän luonnonsuojelun haste.* Helsinki: Into.

Kalliola, Reino. 1958. *Suomen luonto mereltä tuntureille.* Helsinki: WSOY.

Ketola, August. 1955. 'Aulangon villijoutsenet' *Metsästys ja kalastus* 34: 171–173

Koivusaari, Juhani, Ismo Nuuja, Risto Palokangas, Esko Joutsamo, Kaius Hedenström and Torsten Stjernberg. 1973. 'Suomen merikotkat v. 1973'. *Suomen Luonto* 32: 174–77.

Kokko, Yrjö. 1950. *Laulujoutsen – Ultima Thulen lintu.* Helsinki: WSOY.

Lähdesmäki, Heta. 2015. 'Susi yhteiskunnallisena eläimenä'. In Elisa Aaltola and Sámi Kero (eds). *Eläimet yhteiskunnassa.* Helsinki: Into. pp. 185–191.

Lähdesmäki, Heta. 2020. *Susien paikat: Ihminen ja susi 1900-luvun Suomessa.* Turku: Turun yliopisto.

Lehikoinen, Esa, Risto Lemmetyinen, Timo Vuorisalo and Mia Rönkä. 2020. *Suomen lintutiede 1828–1974.* Turku: Faros.

Markus Meier, H.E., Ralf Dötscher and Antti Halkka, 'Simulated distributions of Baltic sea-ice in warming climate and consequences for the winter habitat of the Baltic Ringed Seal'. *Ambio* 33 (2004): 249–56.

Merikallio, Einari. 1950. 'Joutsenen pesiminen Suomessa ennen ja nyt'. *Suomen Luonto* 9: 53–70.

Moilanen, Pekka and Pentti Vikberg (eds). 1986. *Valkohäntäpeura*. Helsinki: Otava.

Mykrä, Sakari, Timo Vuorisalo and Mari Pohja-Mykrä. 2005. 'A history of organized persecution and conservation of wildlife: Species categorizations in Finnish legislation from Medieval times to 1923'. *Oryx* 39.

Nash, Roderick. 1989. *The Rights of Nature: A History of Environmental Ethics*. Madison, WI: University of Wisconsin Press.

Pekurinen, Mika. 1997. 'Sivistys velvoittaa: Klassinen luonnonsuojelu Suomessa'. In Heikki Roiko-Jokela (ed.) *Luonnon ehdoilla vai ihmisen arvoilla? Polemiikkia metsiensuojelusta 1850–1990*. Jyväskylä: Atena. pp. 129–165.

Pohja-Mykrä, Mari. 2014. 'Vahinkoeläinsodasta psykologiseen omistajuuteen: Petokonfliktien historiallinen tausta ja nykypäivän hallinta'. Ph.D. diss. Turku: Turun yliopisto.

Pohja-Mykrä, Mari and Sámi Kurki. 2014. 'Strong community support for illegal killing challenges wolf management'. *European Journal of Wildlife Research* 60: 759–770.

Rantaniemi, A. 1901. 'Muistosanoja majavasta'. *Luonnon ystävä* 5.

Räsänen, Tuomas. 2020. 'Merikotkan ahdinko ja hoivasuojelun synty'. In Tuomas Räsänen and Nora Schuurman (eds). *Kanssakulkijat: Monilajisten kohtaamista jäljillä*. Helsinki: SKS. pp. 285–312.

Robinson, Michael J. 2005. *Predatory Bureaucracy: The Extermination of Wolves and the Transformation of the West*. Boulder, Co.: University of Colorado Press.

Stjernberg, Torsten. 1995. 'Havsörnskydd med avstamp i finsk medeltid'. *Luonnontieteellisen keskusmuseon vuosikirja 1995* Helsinki: Luonnontieteellinen keskusmuseo.

Telkänranta, Helena (ed.) 2008. *Laulujoutsenen perintö: Suomalaisen ympäristöliikkeen taival*. Helsinki: Suomen Luonnonsuojeluliitto.

Teperi, Jouko. 1977. *Sudet Suomen rintamaiden ihmisten uhkana 1800-luvulla*. Helsinki. Suomen Historiallinen Seura.

Tourunen, Auli. 2008. *Animals in an Urban Context: A Zooarchaeological Study of the Medieval and Post-medieval Town of Turku*. Turku: Turun yliopisto.

Uekötter, Frank. 2010. 'Think big: The broad outlines of a burgeoning field'. In Frank Uekötter (ed.) *The Turning Points of Environmental History*. Pittsburgh: University of Pittsburg Press. pp. 3–6.

Vuorisalo, Timo, Esa Lehikoinen and Risto Lemmetyinen. 1999. 'Lintusuojelun varhaisvaiheita Suomessa'. In Timo Soikkanen (ed.) *Ympäristöhistorian näkökulmia: Piispan apajilta trooppiseen helvettiin*. Turku: Turun yliopisto. pp. 104–25.

Vuorisalo, Timo and Markku Oksanen. 2020. '"Mikä on toiselle hyödyksi, voi usein olla toiselle vahingoksi": Pohdintoja eläinluokittelusta'. In In Tuomas Räsänen and Nora Schuurman (eds). *Kanssakulkijat: Monilajisten kohtaamista jäljillä*. Helsinki: SKS. pp. 23–49.

Wallgren, Henrik. 2016. 'Merikotkan onnellinen historia Suomessa'. In Ismo Nuuja and Kalle Ruokolainen (eds). *Merikotkien puolesta: WWF:n merikotkatyöryhmän vuosikymmenten taival*. Helsinki: WWF Suomi. pp. 12–15.

Ylimaunu, Juha. 2000. *Itämeren hylkeenpyyntikulttuurit ja ihminen–hylje -suhde*. Helsinki: SKS.

INDEX

Index

Index

T

U

V

W

X

Y

www.ingramcontent.com/pod-product-compliance
Lightning Source LLC
Chambersburg PA
CBHW020832210326
41598CB00019B/1879

* 9 7 8 1 9 1 2 1 8 6 7 6 1 *